Packaging and Solid Waste
Management Strategies

**Lewis Erwin
L. Hall Healy, Jr.**

A Management Briefing
Developed in Cooperation with
The American Management Association
Packaging Council

Library of Congress Cataloging-in-Publication Data

Erwin, Lewis, 1950–
  Packaging and solid waste : management strategies / Lewis Erwin,
L. Hall Healy, Jr.
    p.   cm.—(AMA management briefing)
  "A management briefing developed in cooperation with the American
Management Association, Packaging Council."
    ISBN 0-8144-2343-4 : $10.00 ($7.50 to members)
    1. Refuse and refuse disposal—United States. 2. Packaging—
Environmental aspects—United States. 3. Waste minimization—
United States. I. Healy, L. Hall, 1941–   . II. Title.
III. Series.
TD788.E78 1989   89-18445
363.72'85—dc20   CIP

© 1990 AMA Membership Publications Division

American Management Association, New York
All rights reserved. Printed in the United States of America.

This Management Briefing has been distributed to all members
enrolled in the American Management Association. Copies may be
purchased at the following single-copy rates: AMA members,
$7.50; Nonmembers, $10.00.

This publication may not be reproduced, stored in a retrieval
system, or transmitted in whole or in part, in any form or by
any means, electronic, mechanical, photocopying, recording, or
otherwise, without the prior written permission of AMA
Membership Publications Division, 135 West 50th Street, New
York, N.Y. 10020.

First printing.

*For the good earth,
and for the wisdom and will to conserve it.*

Acknowledgments

The authors would like to thank the members of the American Management Association Packaging Council. Many members of this group provided invaluable guidance in developing the initial concept, and in reviewing selected segments of the manuscript. (Members of the Council are listed on pages 95 and 96.)

We wish to give special thanks to Council Chairman Bob Esse and to AMA Program Direct Richard Akagi for their encouragement, advice, and support. We would like to thank Don Bohl of AMA and Jacqueline Erwin for their editorial aid, and Judd Alexander for his review and commentary. Finally, we thank our families for their patience and support.

Development and distribution of this publication to AMA members was made possible, in part, by generous contributions from the following companies:

General Mills
R.J. Reynolds Tobacco Company
M&M/Mars

The editors as well as the authors gratefully acknowledge their support.

About the Authors

Lewis Erwin is a professor of mechanical engineering at Northwestern University. Prior to assuming that position, he was the director of packaging technology for Kraft, Inc. He received his PhD from the Massachusetts Institute of Technology and has served on the faculties of MIT and the University of Wisconsin. Dr. Erwin has published more than 60 engineering and technical papers.

Hall Healy is director of marketing at Patrick Engineering, Inc., a full-service civil engineering firm with a significant practice in landfill and related solid waste management work. Previously, he was in new-business product development and strategic planning with two packaging equipment firms, Videojet Systems, International, a subsidiary of A.B. Dick Co., and Signode Corporation, a subsidiary of Illinois Tool Works. Mr. Healy serves on the board of trustees for the Nature Conservancy, Illinois Chapter, and is active in various recycling and environmental organizations. He graduated from Colgate University, received an MBA from the University of Chicago's executive program, and served as a captain in the U.S. Air Force.

Foreword

We who work in the packaging industry as professionals have a deep and real concern regarding the status of solid waste issues in the United States today. These issues affect us personally in our home life and on the job—we live with them all the time. There can be no doubt in anyone's mind that these are extremely complex problems, and that the answers will not come easily.

This document is an attempt to shed some light on the subject as viewed by dedicated people committed to finding solutions within the boundaries of available and emerging technology. It will take much education to get all of us to understand the proper applications of biodegradable plastics or to differentiate between a litter problem and a solid waste problem. We who work full time within this discipline do not always agree on the appropriate solutions.

How can people who are adjacent to this industry or remotely connected as users be expected to intelligently discuss this subject? We hope this document gives some answers by providing both an appropriate level of background information as well as direction in the form of a "management agenda."

Living in an age of information also means living with a quantity of "dis-information." There are statistics available to support any position one might wish to take, and therein lies one of the problems of understanding the solid waste issue. The data given herein have been sifted and verified and only the most reliable sources have been used. The intent is not to overwhelm you with data, but to create an appropriate level of understanding. We hope that this briefing will help you grasp the scope of the issues, how they affect your particular interests, and what can be done to find appropriate long-term solutions.

>Robert L. Esse, Department Head,
>Advanced Packaging Technologies,
>General Mills, and
>Chairman, AMA Packaging Council

Contents

Introduction		11
1	Defining the Problem	13
2	Common Misconceptions	25
3	Methods of Solid Waste Disposal	31
4	Understanding Packaging Solid Waste	40
5	Management Agenda	69
Appendix **A**:	The EPA Agenda for Action	77
Appendix **B**:	General Legislative Framework for Environmental Issues	79
Appendix **C**:	Solid Waste, Packaging-Related University Programs	85
Appendix **D**:	Selected Organizations Dealing with Packaging, Solid Waste Management	89
Additional Reading		93
The Members of the AMA Packaging Council		95

Introduction

Solid waste disposal is now an issue of high visibility in virtually all states. Communities face rapidly increasing disposal costs; opinion leaders denounce the amount of material being thrown away as environmentally and ethically unsound; lawmakers struggle for solutions where few can be found.

This is leading to pressure on manufacturers, particularly packagers of consumer goods, to minimize the amount of refuse. Some jurisdictions have passed laws restricting the types and amounts of packaging permitted, and many others have similar bans on their agendas. If we allow laws based on emotion and poor understanding of environmental and economic impacts to be enacted, we will do a great disservice to the environment, our country, and our businesses.

This book outlines the solid waste crisis, its causes and potential solutions, and describes industry's role in the political and technological management of the crisis.

Our concept of what role industry can and should play draws from the Environmental Protection Agency (EPA) document *The Solid Waste Dilemma: An Agenda for Action*. This publication, summarized in Appendix A, is required reading for anyone concerned with working toward long-term solutions. In essence, the EPA's *Agenda for Action* establishes broad priorities: first, source reduction, then recycling, then incineration, and finally, landfill. We believe that this hierarchy provides a reasonable and effective framework for both lawmakers and industry.

Our discussion focuses exclusively on issues related to municipal solid waste (as opposed to hazardous waste or industrial waste), with particular attention to the position of packaging in this waste stream. Municipal solid waste is made up of refuse from residences, institutions, and commercial establishments, as well as from public works (street cleaning, for example). It includes waste from some industry sources (corrugated boxes and cafeteria wastes, for example), but not industrial process waste.

The statistics presented here use weight as the measure of significance, for reasons explained later. Although there are situations in which compacted volume is also a significant measure, reliable data on compacted volume are not currently available.

This book is addressed to managers in companies who use, handle or sell packaging, not just to the food and beverage industry. The current visibility of the solid waste problem in our society is creating legislation that will impact all users and handlers of packaging, industrial as well as consumers.

1

Defining the Problem

The United States does not have a *national* solid waste problem. There is, however, a nationwide epidemic of local solid waste problems. If our waste were discarded on a national or international basis, as our coal is mined and our oil is pumped, there would be no solid waste crisis. Clearly, there would still be environmental impact associated with solid waste, but this would be minor compared with the impact of coal, oil, and other extractive industries. However, our society views trash as a local issue and it is likely to stay that way.

Thus, there is not one solid waste problem, but many. We have a northern New Jersey solid waste problem, a Seattle solid waste problem, a Chicago solid waste problem, and so on, across the nation. Each of these locations will work toward solutions appropriate to its own geography, population density, and prevailing lifestyles. Each will be largely unconcerned with the solutions other localities are using unless such practices affect it locally.

This purely local point of view will cause problems for national marketers of goods. State and local governments propose packaging regulations based on what they believe will support most beneficially their own solid waste disposal

---

Digging Holes, Filling Holes
A Sense of Perspective

Americans produced 160 million tons of municipal solid waste in 1988.[1] Disposing of this in landfills and incinerators at an average cost of $26 per ton amounts to a total of nearly $4.2 million, or about $18 per person. One hundred sixty million tons appears to be a great amount of material to handle, but this quantity is not large by the standards of our society. We dig up 800 million tons of coal and 1,000 million tons of sand and gravel each year. We are making holes in the ground far faster than we are filling them.

Solid waste is not the major material handling effort of our society, nor is landfill the major cause of environmentally disruptive use of land area. *(Information compiled by Judd Alexander of James River Corp.)*

---

systems. Different localities have different disposal systems; decision making in each is susceptible to pressure from special interest groups. As a result, the requirements will vary.

Such a checkerboard pattern threatens the efficient distribution of goods in our national economy and escalates the cost of goods, as package designers struggle to comply with local codes that require inefficient packaging. Some container suppliers suffer because their packages are affected (or are feared to be affected) by the regulations.

Companies that do not respond to what is perceived as a crucial and pervasive need will be cast as irresponsible.

At the same time, local governments may be no closer to long-term solutions than before they enacted the laws. Many of the codes currently under consideration are aimed at superficial factors rather than root causes.

The tendency to adopt superficial solutions is, in part, a result of frustration. The problems are incredibly complex; comprehensive solutions seem elusive. Yet, something must be done (see box, pages 15 to 18).

Are comprehensive solutions possible? Before answering that question, let's look at the situation as it currently exists.

(Continued, page 19)

How State Lawmakers See the Issues

To gain insight into how state lawmakers view the problems and issues, AMA researchers talked with three legislators in the forefront of solid waste management. All three represent states in the high-density Northeast, where the problems are most severe. What is happening in these crowded states now will influence action in other states shortly.

The three representatives interviewed were:

- Representative Mary Mushinsky, Chairperson of the Environment Committee of the Connecticut General Assembly.
- Assemblyman Maurice Hinchey, Chairperson of the Legislative Committee on Solid Waste Management, New York.
- Representative Robert Shinn, Jr., Chairperson of the Assembly Solid Waste Management Committee, New Jersey.

A consensus was readily apparent on several points:

- Time has run out. We are fast approaching—or have already reached—the crisis stage. States can no longer handle the waste being generated, and disposal costs are escalating at a rapid pace.
- Solutions must come in three areas: source reduction, recycling, and new and/or increased disposal capacity—incinerators, landfills, etc.
- Of these solutions, only source reduction doesn't require substantial municipal or state funding.
- Recycling is an important part of all plans. These states, along with others, have already passed mandatory recycling bills.
- Opening new landfills is, without doubt, the most difficult process. Public outcry is loud and strong.

Essentially, the state legislatures are looking to industry for significant reductions in waste material produced and for substantially higher levels of recyclable products and packaging. They are hoping for cooperation and support in this effort but are prepared to force the issue if necessary. All are using a "carrot-and-stick" (tax credits and taxes) approach to one degree or another.

However, "The carrots are not working all that well," as Representative Shinn put it. Assemblyman Hinchey voiced the same concern: "The more progressive businesses—the smart ones—will take action now. The others . . . well, they will have to be told." Representative Mushinsky underscored the point. "The crowded states are really getting hard-nosed about this [source reduction and recyclability], because they are desperate."

The problem, simply defined, is that new disposal capacity cannot keep pace with the increasing volume of waste. Thus, source reduction becomes the top priority on everyone's list. As Mushinsky points out, "We are looking at a 10 to 20 percent increase in our rate of generation of solid waste in the next eleven years. We have got to get that growth rate under control." Since packaging accounts for approximately one-third by weight (50 percent by volume, according to Mushinsky) of the total waste stream, it is understandable that packagers are prime targets in any source reduction campaign.

When it comes to opening a new landfill, nobody wants it in their neighborhood. Even dedicated environmentalists suggest alternative locations that would be "better." This "somewhere else" approach to undesirable intrusions has acquired the acronym NIMBY—not in MY backyard.

New landfills are needed near the crowded metropolitan areas where the waste is being generated. They can be located least intrusively in low-density rural areas. But low-density rural areas are virtually nonexistent in Connecticut and New Jersey, and are disappearing fast throughout the other high-population states.

Consequently, a deafening hue and cry is raised by citizens (voters) whenever a new waste facility site is proposed. Representative Mushinsky described the problem:

> When we have to pick a site for a waste plant or transfer station, it's a miserable job. The towns do everything in their power to defeat us. We have tried to build [facilities] and we get tied up in [law] suits, and public officials get defeated in the next election.

Representative Shinn said it best:

> There never was a harder job. There are three problems: siting, siting, and siting. Municipal site selectors are the closest thing we have to kamikaze pilots in this country.

Transfer stations and incinerators are not as hard to site as landfills, but they are still not easy, according to all three representatives.

But something has to be done.

Thus, attention turns to the alternatives: source reduction and recycling. Both are welcomed by environmentalists and neither requires an unwanted intrusion. There are no unsightly emissions or leaching problems that could jeopardize groundwater. Increased truck traffic to a recycling facility may be a problem, but it is minimal in comparison.

Source reduction is the simplest and most basic of all. If packages were reduced by half, said the representatives, marketers and manufacturers might complain, but there would be no complaints from the voters. With tipping fees as high as $150 a ton (and going higher) on the East Coast, source reduction takes on still greater urgency.

Recycling is also appealing, and New Jersey, Connecticut, and New York have passed stringent recycling laws, as have many other states.

In New Jersey, source separation by residential customers is now mandatory throughout the state. In New York, the law requires each municipality to establish its own regulations for separating out paper, plastics, metal, and glass by September 1992. While the initial emphasis is on residential recycling, mandatory commercial recycling is not far off. (It is already a reality in Rhode Island.)

Connecticut law stipulates that by January 1991 every household must have two trash cans, one for all mixed recyclables and the other for everything else. The mixed recyclables will be shipped to regional reprocessing centers for separation. Businesses are included under the Connecticut law and they must separate out white paper and corrugated cardboard as well.

WHAT LEGISLATORS WANT FROM INDUSTRY

When legislators look at the content of the waste stream, they do not see packaging as a factor in sales, and they are simply not worried that different regulations in each state can create havoc with product production. As Assemblyman Hinchey put it, "So much packaging would seem to be unnecessary. So regulating it, reducing it, and taxing it does not seem unfair."

Growth in recycling requires corresponding growth in the market for recycled products. Consequently, when states look to business for participation in recycling, they are also looking to business to provide a market. As Ms. Mushinsky notes, "A lot of companies still make virgin paper boxes. That is totally unnecessary."

Moreover, the states are looking for what Ms. Mushinsky describes as "recycle friendly" products and packages. One key aspect of this is packaging that is not made up of more than one recyclable material. Combinations of paper and foil and plastic are difficult to separate out even though each may be a recyclable element.

In terms of recycling, some industries (such as aluminum) are well advanced, and others are lagging. Mushinsky probably expresses the attitude of many legislators when she notes, "If the plastics industry put its mind to it, it could develop a technology to make recycling practical. In fact, the plastics industry is going to have to do just that if it wants to survive, because we are just one of the early states to pick on them. Every state on the East Coast and California on the West Coast is in the same boat. It is just that we got there first."

In essence, the state legislators are saying to businesses: This is a serious problem, but we are going to solve it. It will take sacrifices on everybody's part, including industry. If you are going to join us in developing these new rules, which will hurt, then we want to work with you. If you don't cooperate—really cooperate—then we will make the rules and you will have to figure out for yourselves how to live with them. The new rules have to come whether you like it or not.

HOW WE ARRIVED AT OUR CURRENT SITUATION

Viewed from a very basic level, the problems are both economic and political. We have created a vigorous economy that produces solid waste in increasing amounts as the economy grows without providing a business infrastructure and a political will for responsibly disposing of the waste. Exhibit 1.1 shows the rise in solid waste between 1970 and 1986. All types of solid waste increased in that period.

The failure to find comprehensive solutions has a number of significant consequences.

1. Communities have fallen back on simple, one-method disposal systems. Today, 80 percent of the nation's solid waste is landfilled, 10 percent is recycled, and 10 percent is incinerated.[2] The preponderance of landfilling demonstrates the appeal of a one-method system. It can be managed simply. Mass-burn incinerators also have such appeal.

2. It is becoming increasingly difficult to operate in this context. One-third of our landfills will be full by 1993.[3] New disposal systems will need to come on line to replace these. Many existing facilities are closing because they do not meet

Exhibit 1.1. Growth in municipal solid waste, 1970–1986.

Population	+	18%
MSW	+	25%
MSW Components:		
Packaging	+	9%
Durables	+	35%
Newspapers/Magazines/Books	+	40%
Office/Commerical and Other Paper	+	69%
Misc. Non-Durables	+	300%
Clothing/Footwear	+	88%

Source: *Characteristics of Municipal Solid Waste in the United States, Update 1988,* Franklin Associates, Ltd. *Analysis by Judd Alexander.*

Packaging and Solid Waste—19

state or federal standards for protection of human health or the environment.

New disposal facilities are politically difficult to open. They are opposed for environmental reasons, especially perceived potential ground and surface water pollution and air emissions, as well as nuisance factors such as truck traffic and anxiety over property values. Our local governments have not developed conflict-resolution mechanisms to enable timely and appropriate siting of refuse disposal facilities.

3. To complicate matters further, society's expectations related to environmental protection continue to rise. Methods used even a decade ago for disposing of trash are unacceptable today.

Until the 1960s, much municipal solid waste was disposed of by environmentally abusive methods: open dumps and crude incinerators. Open dumps were simple pits into which refuse was discarded. The mass of the material was reduced by vermin eating the biodegradable portions and by periodic burning, accompanied by air pollution. The remaining material slowly filled up the pit. Open dumps lasted a long time. Crude incinerators reduced the mass of the waste by combustion, emitting offensive smoke and often other pollutants in the process.

When the abusiveness of these methods was recognized, they were replaced by sanitary landfills and modern incinerators. Landfills fill up more rapidly than old dumps because there is much slower decay and no burning. This requires more frequent establishment of new sites. New sites are becoming expensive in dollars and in political goodwill as people become aware of the environment and of their ability to prevent the siting of what they perceive to be objectionable facilities in their neighborhoods.

4. Still further complications arise because of misconceptions regarding solid waste. Consumers assess the extent of the problem according to the number of bags they place on the curb or the quantity of litter they observe. These intuitive measures are often not the best gauges of environmental impact. Trash on the curb represents uncompacted volume; litter and solid waste are actually different problems (see

Chapter 2). Misunderstandings such as these lead to conflict in the policy-making process.

ENTER, SPECIAL INTEREST GROUPS

It is within this context that various special interest groups begin to exert influence. The diversity in points of view leads to conflicting proposals for solutions.

> MUNICIPAL GOVERNMENTS carry disposal costs. They are responsible for getting rid of solid waste. They must either bury it, burn it, recycle it, or get people to quit generating it. If that means getting manufacturers to stop supplying it, they include such constraints in their local regulatory agenda.
>
> ENVIRONMENTALISTS tend to focus on particular issues such as litter, air or water pollution, or natural resource conservation and human lifestyle. Some maintain that a throw-away lifestyle is intrinsically evil. Others are concerned about marine life and target packaging discarded in marine environments.
>
> COMMERCIAL RECYCLERS, including scrap dealers, have a vested business interest in recycling and are often viewed as environmental experts by those who wish to foster recycling. While recycling is a key part of an integrated waste management system, overemphasis on recycling can inhibit source reduction.

Segments of the PACKAGING SUPPLY INDUSTRY have sought to use the solid waste issue to gain competitive advantage for their material.

> The ALUMINUM industry has told its story well. Fifty percent of aluminum cans are recycled, and as a result this industry is perceived as a hero by environmentalists.
>
> The GLASS industry is also telling its story. It has created an infrastructure to reuse consumer scrap glass and is using recyclability (current estimates range from 17 to 27 percent)[4] to preserve its share of the rigid container market.
>
> The PLASTIC industry has so far done a less aggressive job of telling its story and creating a recycling network. Plastic is still

popularly viewed as a negative factor in the solid waste crisis. Extensive new efforts are underway, but their effectiveness has yet to be demonstrated in a manner to sway significant public opinion.

The BEVERAGE industries are sensitive to the litter issue. They are impacted by deposit laws. Essentially all recycled glass containers are beverage glass.

The FAST FOOD industry is also sensitive to the litter issue. It has modified packaging and set up recycling programs.

The GROCERY FOOD industry has been politically quiet.

OTHER INDUSTRIES (all of which use packaging) have been quiet and may not even realize that packaging laws will affect them.

Legislators seek to develop state programs to respond to problems they see in their municipalities. The special interest lobbies have been exceptionally active at this level. As a result, legislative actions are often well-intended but ill-advised: bans on nonrecyclable materials and plastic containers, more deposit laws, and the like. Beverage containers remain the primary target, but food packages (indeed all packaging) are being included in many regulations.

TOWARD A COMPREHENSIVE SOLUTION

These problems will not go away easily. Most communities continue to rely heavily on landfills and incinerators—organizationally simple solutions—at the expense of *integrated waste management*, a more environmentally sound but more management-intensive approach.

Of all that has been formulated to date, the concepts of integrated waste management (as articulated in the *EPA Agenda for Action*) appear to offer the best foundation for long-term, environmentally responsible solutions. Building such systems, however, will demand understanding and cooperation from both industry and government.

The following chapters are intended to build a foundation for that understanding and cooperation.

1. The next chapter attempts to correct some of the major misconceptions regarding solid waste. Only when such misunderstandings are removed can we begin to see our way clearly toward long-term solutions. In addition, these clarifications add greater credibility to the EPA *Agenda for Action*.

2. Chapter 3 provides information on the major elements of the EPA hierarchy—source reduction, recycling, incineration, landfill. Where do we currently stand with regard to technologies and infrastructures needed to act on this hierarchy?

3. Chapter 4 provides extensive information on packaging materials as components of the solid waste stream. We have attempted to provide objective, statistical information, without bias toward any particular material.

4. Based on this understanding, how can industry begin to work toward long-term solutions? Chapter 5 suggests an agenda.

References

1. *The Solid Waste Dilemma: An Agenda for Action*, U.S. Environmental Protection Agency, 1989 (EPA/530–sw–89–019), p. 1 (hereafter referred to as *EPA: Agenda for Action*).
2. *EPA: Agenda for Action*.
3. *EPA: Agenda for Action*, p. 14.
4. U.S. Department of Commerce.

2

Common Misconceptions

A key aspect of the solid waste problem is misinformation. Common misunderstandings on the public's part, combined with the emotional aspect of the problem, are the impetus behind inappropriate legislation. The solid waste crisis does not have simple saviors or demons. An examination of the most fervently believed misconceptions will demonstrate this.

"The solid waste problem is obvious; just look at all that litter along the road."

Litter and solid waste are actually very different problems; they require different solutions.

Consider, for example, the impact of mandatory deposit legislation ("bottle bills"). These measures can dramatically reduce litter; at the same time, they have an indirect impact on solid waste by encouraging segregated (and thereby more readily recyclable) waste components. In some cases, the deposit legislation may actually inhibit more comprehensive recycling. On the other hand, centers for voluntary recycling of glass bottles reduce the amount of solid waste going to landfills but have little or no impact on litter.

Litter comes from improper disposal of solid waste. It is unsightly and expensive to pick up, but small in tonnage compared with the greater problem. Solutions to the solid waste problem require a focus on the handling of material that *is* properly disposed of.

"We can solve the solid waste problem by getting rid of all that excess packaging."

Packaging represents only 33 percent of solid waste. Thus, a reduction in packaging, by itself, will not solve the problem. In fact, packaging has been steadily reduced. The annual amount of packaging waste per person has been declining over the past 20 years.

This does not mean, however, that further reductions are not possible or desirable. Two principles must be kept in mind as we work toward this goal.

First, packaging exists for both consumer appeal and good product protection. No one can argue that product protection should be compromised.

Second, reduction in packaging is of far greater benefit to the packager than to the disposal system. That is, if we consider the entire life-cycle costs of a package (raw materials, energy used in manufacture, transportation costs, disposal costs, and the like), disposal is a very small part of the total equation (typically a few percentage points). On the other hand, the packager's costs can be considerable.

Thus, a very strong motivation already exists for the packager to reduce energy and raw materials.

"I know how much solid waste I make; I count the number of trash cans and bags I put on the curb."

Actually, the most appropriate measure of the environmental and economic impact of waste depends on the disposal technique. When transportation is the primary cost, total weight is the relevant measure. This is also the measure of cost impact at many incinerators and landfills, since tipping fees are set on a weight basis. The environmental impact at landfills is best measured by compacted volume, as this determines how fast the landfill fills up.

Consumers assess the quantity of solid waste by the number of cans or bags they place on the curb. This represents uncompacted volume, a measure that does not give an accurate reading of either cost or environmental impact.

"Glass is environmentally superior to plastic."

The fact is that plastic is not intrinsically more harmful than any other packaging material.

This point was underscored by the United States Bureau of Alcohol, Tobacco and Firearms in the process of authorizing plastic liquor bottles. In developing an environmental impact statement comparing glass and plastic, the bureau considered each material's use of natural resources (particularly petroleum and natural gas) as well as the impact of discarded containers on solid waste. It found no environmental harm in the conversion from glass to plastic.

"Refillable containers are always environmentally more benign than one-way containers."

To withstand the process of return and refill, a refillable container must be about twice as heavy as a one-way container of the same material. This means that about twice as much oil and natural gas are required for the manufacture and distribution of the container, and the disposal generates twice as much waste.

The doubling of the costs can, of course, be amortized over a greater number of trips. The more trips, the lower the environmental impact per trip.

The number of trips required for breaking even depends on the distance to the bottling plant, since transportation energy is a significant component of the environmental impact. In situations where a very high return rate is achieved and the bottling plant is close to the users, refillables can reduce the environmental impact of beverage use. In many places, however, the number of trips is not sufficient and the distance to the bottling plant is too great, meaning that refillable bottles can actually impose a more severe environmental impact than one-way bottles.

"Packaging is hurting the ozone layer."

Current consumer packaging materials contain none of the chlorofluorocarbon (CFC) materials identified with damage to the ozone layer. In the past, some types of packaging foams were made using CFC. Less harmful gasses have been substituted. Small amounts of CFC are still used in some *industrial* packaging where suitable replacements have not yet been developed.

"Things made of recycled materials are inferior to goods made with new materials."

Recycled aluminum, glass, and steel are used interchangeably with new materials. Recycled paper, however, must be used with care, as the fiber length is shortened and it contains ink, which prevents achieving a bright white color. Recycled plastic may suffer some degradation, depending on the plastic and the manufacturing process used. Plastics-molding plants routinely use the scrap from defective parts, trimmings, and similar internal sources, blending it into new material. Postconsumer plastic scrap is used interchangeably with new materials in the manufacture of fiber-fill insulation, strapping, and geotextiles.

The quality of recycled material depends primarily on the care used in separating and cleaning the material. Sufficiently clean recycled material is competitive in quality with new as a raw material.

"Biodegradable packaging will solve the solid waste problem."

The National Parks System tells us "Take only pictures; leave only footprints." This is a laudable slogan for minimizing the environmental impact of our lifestyles. We would all like our packaging to disappear as easily as footprints in the national parks. If soft drink cups or potato chip bags could be tossed out the window to be eaten by a passing raccoon or songbird, we would feel much more positive about packaging.

Simplistic thinking such as this will not help us achieve environmental goals. Biodegradable packaging is one example of a solution with great superficial appeal. Few are cognizant of its underlying potential for harm. Because of the controversy surrounding this innovation, we will look at it in greater detail than the other misconceptions.

A level of degradability in which packages disappear as easily as a potato skin or an apple peel simply cannot be achieved with today's technology. We aspire to it in research, but we must not overpromise the results of biodegradable packaging and should not pass laws requiring a technology that does not exist.

Polyethylene (the most common plastic bag material) can be rendered biodegradable by blending it with cornstarch. Let us examine this action on the solid waste system by reviewing the impact on each of the strategies in the *EPA Agenda for Action*.

Source Reduction (Reducing the Amount of Material Entering the Waste Stream)

Biodegradable polyethylene is weaker than unmodified polyethylene. As a result, biodegradable bags must be thicker and heavier than conventional bags. Such a change increases the amount of material in a bag—and the amount of material discarded. Requiring biodegradability defeats source reduction.

Recycling

Requiring some polyethylene to be biodegradable complicates the recycling process in two ways. By adding new material to the waste stream, it complicates separation of plastics for reuse. It also makes the products produced from recycled polyethylene biodegradable, limiting the applications for the recycled material.

Incineration

Biodegradable materials absorb water. This makes them more difficult to burn and decreases the amount of recoverable energy produced by burning. Requiring biodegradability undermines incineration.

Landfill

Biologically degrading materials cause most of the environmental impact of municipal solid waste landfills. If a landfill has only nondegrading, insoluble solids, it will not cause groundwater contamination, will not have major surface subsidence, and will not generate methane gas. Requiring biodegradability makes landfills more environmentally harmful.

Thus, in each major strategy for dealing with solid waste, required biodegradability increases the problem. Clearly, it is not "the solution" to the solid waste problem.

It is equally clear, however, that in some applications biodegradable packaging is environmentally advantageous.

Trash that is discarded into water should degrade in water. Waste thrown overboard by ships or flushed into municipal wastewater floats about the ocean until it is washed up on shore and buried by wave action. Such waste has a long life as litter pollution. In addition to being ugly, it can harm fish, mammals, and birds if ingested. Such techniques of solid waste disposal are environmentally reprehensible but can be alleviated somewhat through the use of biodegradable materials.

Containers for waste that is composted should be made of materials compatible with that process. Leaf bags in communities that compost their leaves should be made of degradable materials.

It is advantageous if items with a particular propensity to appear as litter are made of biodegradable materials. However, the current rate of biodegradability is so slow that in any environment where trash is cleaned up it will be collected before it disappears through degradation. If consumers exhibit a greater tendency to litter because they believe biodegradable trash is less environmentally harmful, we will increase, not decrease, our litter problem.

"We've achieved recycling by requiring a return deposit on all beverage bottles."

Seven states have set mandatory deposits on beverage bottles. These bottle bills hark back to the days of refillable bottles and carry the nostalgic appeal of an era when children collected litter to buy bubble gum. The purpose of the laws is to encourage proper disposal and to reward litter collectors. As litter suppression methods, they are effective—but expensive.

Their effect on solid waste is more complex in that they provide a partial (rather than a comprehensive) basis for recycling.

Deposit bottles are redeemed at food and beverage stores or at special redemption centers. This results in a large amount of well-segregated, uniform solid waste. While some of these bottles are sent to bottlers for cleaning and refilling, most are not. The weight and cost penalty of manufacturing bottles that can be refilled combined with high labor and transportation costs makes the whole process too expensive.

The nonrefillable bottles are sold to materials recyclers who crush and clean the glass, or grind and segregate the plastic. (Most of the plastic bottles recycled from postconsumer use in 1988 came from this process.)

Thus, bottle bills have had a significant impact in promoting the development of a recycling infrastructure and a market for postconsumer scrap; at the same time, they have been effective in reducing litter.

In addition to adding expense for both consumers and retailers, the system has one very severe limitation. It skims the high-value recyclables off the waste stream without establishing a system that can be expanded to include additional, less valuable materials. The valuable recyclables in the waste stream are paper (newspapers and corrugated board), glass, aluminum, and certain plastics, when segregated by type. The principal sources of such plastic, aluminum, and glass are beverage bottles and cans. When these are removed from the waste stream, much of the potential profitability of materials recovery is also removed. This inhibits the development of comprehensive recovery systems.

3

Methods of Solid Waste Disposal

The less one has to deal with, the easier the disposal. Of all the methods for confronting the solid waste crisis, source reduction is the most appealing—as well as the least well-defined or understood.

Simply stated, source reduction is reducing the *amount* or the *toxicity* of the material entering the solid waste stream.

To some degree, the goals of source reduction are at odds with society's goals of economic growth. Everything manufactured becomes waste. If we wish to maintain a high standard of living while lessening the waste created, we must deliver the same or better goods and services using less material.

Source reduction includes both lessening the amount of solid waste and reducing the waste's toxicity. In the packaging area, the first objective is accomplished through lightweighting packages. In general, American industry has already made marked progress in this regard.

Recall, from Exhibit 1.1, that all components in the municipal solid waste stream increased from 1970 to 1986.

A closer look at that exhibit reveals that the increase in packaging was less than the increase in population, signaling a decrease in the amount of per-capita packaging. The reduction has come not as a result of consumers buying fewer packaged goods nor from the use of inferior packaging to minimize waste. Rather, the credit goes to packaging engineers who have reduced the amount of material used at the same time that they have worked to make packages more convenient. Actually, the driving force behind these gains has been cost rather than disposal issues. Exhibit 3.1 shows the cost of several packaging materials on a per-ton basis. Comparing, for example, the cost per ton that a company pays for plastic bottles with the $25 per ton saved in avoided landfill costs underscores a point made in the previous chapter: the major savings in source reduction accrue to the packaging company. Of all the parties involved, the company has the strongest motivation for reduction.

In general, the public has not been made aware of the gains made thus far. Indeed, it is the goal of the packaging engineer to produce a less expensive package (a lighter one, using less material) that performs in a manner identical to

Exhibit 3.1. Source reduction is cost driven.

Initial Cost Is a Very Strong Driving Force for Minimizing Amount of Packaging

Landfill Cost	$10 to $140/Ton
	National Average $26.50/Ton

Packaging Material Cost to Food Company

Foil	$6000/Ton
Paper Cartons	$ 800 to 1800/Ton
Plastic Wrap	$2500 to 4000/Ton
Plastic Bottles & Cups	$1400 to 4000/Ton
Glass Bottles	$ 200 to 400/Ton

the conventional package, and to do this in such a way that the consumer does not see the difference.

Some legislative and environmental agendas have included source-reduction proposals that have been developed without benefit of the packaging engineer's knowledge of what is effective, or without a view of the total environmental impact. These proposals include the required use of refillable bottles, deletion of protective cartons on tubes, elimination of the paper sleeve on gum wrappers, deletion of wrapping films on cartons, banning "individual portion" packages of condiments, and even requiring consumers to supply their own containers for bringing food home from the supermarket.

Most of these proposals have adverse economic, health, or environmental impact. They are proposed out of frustration and without an understanding that source reduction is already being accomplished. The advantage of such draconian measures is that the impact can be seen. Unfortunately, the benefits are small compared with ongoing source reduction driven by cost reduction.

American industry must develop measures to demonstrate the impact of ongoing source reduction so that the pressures for such ill-advised legislation will lessen.

Recycling

When a material cannot be prevented from entering the waste stream, the preferred method of dealing with it is recycling. In a sense, this is the most "natural" method of handling waste. Throughout the plant and animal kingdoms, one organism's waste is another's food. Natural selection has provided biological organisms that occupy niches created by the products of other organisms. This process is too slow, however, to deal with the rapid development of new waste materials created by human technology. Since glass and fired clay still have no biological degradation mechanisms (even though they have existed for thousands of years), it should not surprise us that plastic and aluminum also have none.

In the absence of natural recyclers, the technological system that developed the materials must also develop the niche for reuse of the waste materials.

Today, about 10 percent of the solid waste in the United States is recycled. The EPA has set a goal of recycling 25 percent by 1992.[1] Achieving this goal will require major efforts.

When the economic and environmental costs of landfill and incineration increase in magnitude and visibility, activists bring pressure on the sources of solid waste. The basic materials industries have responded with scrap buy-back programs for recycling. The aluminum, glass, and paper industries have had such programs for decades. The plastics industry has begun to develop such a system only recently.

The recyclability of a material is determined by the existence of a market for the material.

To recycle a material, it must first be segregated from the waste stream, then processed into a usable form, and finally sold to a user. These three steps require an infrastructure of institutions and practices in municipalities, secondary materials processors and brokers, and basic materials processors. This infrastructure is complex; growth among the various institutions needs to be coordinated. The recyclers will not develop practices and processes for recycling unless the market for the material is reliable. And users of recycled materials will not provide a reliable market unless there is assurance of stability in quantity and quality. Building such systems takes time and money.

From the consumer's point of view, recycling is separating bottles, cans, and paper and taking these to the recycling center. Industries that wish to be viewed as environmentally responsible want their materials represented among the bins at the recycling center. The bins are provided for materials for which there is a market. This creates a political benefit that pressures basic materials industries to foster the development of an infrastructure, which in turn creates a larger market for their materials.

Recycling has two benefits: It is a source of raw material, and it avoids landfill/incineration. Both benefits carry an economic value.

The value of material segregated for recycling will depend on the cost of processing and transportation as well as on the value of the product produced from the recycled material. The value of landfill/incineration avoidance will depend on the local cost of such operations.

This means that in remote areas, where landfill is cheap and transportation costs to recycling plants are high, relatively few materials will be "recyclable." In areas of high population density, where manufacturers (the users of recycled materials) are near and where landfill real estate is expensive, many more materials will be recyclable. Thus, recyclability is determined by the geographical and industrial environment as well as by the material.

The economics of recycling common packaging materials is worth examination.

Aluminum is the most recyclable packaging material primarily because the manufacture of aluminum metal is energy-intensive. This energy is not required in processing scrap, thus creating a powerful economic incentive for recycling. Reprocessing aluminum is easy and safe. The high temperature needed to melt the metal kills microbial contaminants and destroys any organic toxins. The low viscosity of the molten aluminum allows other contamination to be removed by flotation of sedimentation. In short, recycled aluminum is a clean material of high value, and this value has become the dominant incentive for recycling.

Glass can also be melted and reprocessed into new containers. However, the energy savings are not as dramatic as they are for aluminum. Unlike aluminum, the energy of melting the glass is the major energy requirement of producing the container. Recycling glass saves sand and other materials and uses approximately 15 percent less energy than does new manufacture. This energy saving is real, but can be wiped out by the transportation energy required to bring the recycled material to the glass plant. If the scrap must be trucked more than a few hundred miles, the recycled glass will actually require more energy.

Because glass containers are heavy and cannot be incinerated, landfill avoidance will be a major incentive for recycling. The glass industry, sensitive to the environmental impact of glass, has developed a recycling infrastructure which ensures that communities that separate out their glass waste will have a market for the material. This assurance, combined with the recycling ethic, has made glass recycling a fixture of American waste management.

Paper has also been recycled for generations. When paper is available in uniform quality and quantity, it is reprocessed readily into paper and paperboard products. In the consumer waste stream, newspapers are typically the most common form of paper captured for recycling. In the industrial waste stream, corrugated boxes, captured at the point where they are opened (supermarkets, for example), are the most common form.

The key issues in recycling paper are ink removal and fiber-length degradation. Ink must be removed if a light-color paper is to be obtained. When paper is repulped for reforming, the action shortens the length of the cellulose fibers, lessening the paper's strength. Recycled paper does not have the strength or color of paper made from new fibers, but lower cost can make it attractive for many applications.

Incineration

Incineration is a major method for reducing the mass of solid waste. Modern incinerators also produce steam for heating or electrical generation with the energy from burning trash.

Incineration does not eliminate solid waste but can dramatically reduce the volume and weight of waste that must be landfilled. Paper, plastic, food refuse, and yard wastes can be burned. Metal and glass should be separated from the waste stream before being incinerated.

The ash from incinerators contains the unburnable constituents of the solid waste stream and is therefore enriched in heavy metals. These metals, which come primarily from batteries and pigments, are toxic. The concentrations found in incinerator ash are high enough to force the classification of the ash as toxic waste, which greatly increases the cost of disposal. This is not because incineration introduces new toxic materials. Rather, the nontoxic materials have been burned off, resulting in higher concentrations of the toxic substances. A landfill that can legally accept municipal solid waste may not be able to accept the ash from incinerated municipal solid waste.

Modern incinerators have been demonstrated to be safe and clean when properly operated. However, there is a reasonable concern that improperly operated incinerators will produce objectionable or harmful emissions. This, combined with the public's memory of the old technology of less clean incinerators, makes incineration a controversial topic.

It has been suggested that some materials constitute problems in incinerators. In Japan, where incineration has been practiced for decades, many of the older plants are not designed to use the high heat value of plastics. In these areas, plastics are separated out and landfilled.

Materials containing organic chloride produce hydrochloric acid on combustion, and this can corrode older incinerators. They also produce trace amounts of other toxic materials when burned in an improperly operated incinerator. For these reasons, polymers containing chloride (particularly the most common, PVC) have come under pressure.

Landfill

The least desirable but most practical method of waste disposal is landfilling, which currently accounts for 80 percent of the waste disposed.

In a sanitary landfill, the newly discarded refuse is covered each day with six inches of soil. This controls vermin infestation and prevents fires. When the landfill is closed, it is covered with clay to restrict rain water from seeping through the refuse, leaching pollutants into groundwater. These procedures create a dry, low-oxygen environment. In such environments, aerobic degradation mechanisms (fire, rats, aerobic microbes) do not function; thus degradation is very slow. Excavations of landfills reveal twenty-year-old newspapers that are still readable and carrots that are still orange when broken open. Sanitary landfilling, when done properly, is entombment, not composting. Because the more rapid aerobic degradation mechanisms cannot function, landfills are filling up much faster than dumps. This means that new landfill sites must be developed frequently.

A new landfill site should meet three criteria:

1. It must be located in a geological structure formed in such a way that percolating rainwater does not leach pollutants into the groundwater.
2. It should not be surrounded by a population that objects to the site being established.
3. It should be close to the source of solid waste to minimize transportation costs.

Points 2 and 3 are in conflict. Municipal solid waste is created by people in the course of living, eating, drinking, working, and playing. This means that solid waste is created most intensively in regions of high population, particularly major cities. Yet, these are the most difficult places in which to site waste disposal facilities, both because land is expensive and because large numbers of people will usually raise objections. While siting can be compelled, the process is slow and expensive, both in dollars and political favor. Few politicians get votes from neighborhoods where they have sited waste disposal facilities.

Integrated Waste Management

Landfilling and incinerating are currently the two major technological approaches to handling solid waste. Other techniques are also used in some combination with one or both of these. The complementary use of a variety of waste management practices is known as integrated waste management. The methods now employed are not all equally desirable. They can best be thought of as a hierarchy[2]:

- Source reduction
- Recycling (including composting)
- Waste combustion, with energy recovery
- Landfilling

These techniques can be used together to handle solid waste economically while minimizing the environmental impact. Different communities will use these different techniques in ways appropriate to their geography, population density, and waste stream constitution. In isolated communities where land is relatively plentiful and inexpensive and distances to markets for recycled materials are great, landfilling will continue to play the dominant role. In the densely populated regions, other techniques will receive more attention to minimize the need for landfill.

In effective integrated waste management, each of the techniques must be designed to be part of a whole system. For example, combustors used in incineration must be designed to handle the heating value of a previously sorted waste stream, since some high heating value components will have been removed for recycling as materials and will be unavailable for fuel.

The next chapter takes a closer look at how each of the techniques affects various materials in the solid waste stream.

References

1. *EPA: Agenda for Action.*
2. The Garbage Project, Bureau of Applied Research in Anthropology, University of Arizona, *The Mullins Dig: An Archaeological Evacuation of Three Modern Landfills.*

4

Understanding Packaging Solid Waste

People have had to deal with waste from prehistoric times to the modern era. Mankind has dumped, burned, or reused refuse in both primitive and civilized societies. By studying historic dump sites, we have learned a great deal about civilizations—even our own.

In ancient times, Romans took their refuse far enough out of town to keep from smelling it. The American Plains Indian was an early judicious recycler, carefully reusing all parts of the buffalo. Eskimos did the same with the whale. Colonists recycled fabrics by making quilts. Like the old knife-sharpening vendor wandering down early streets, scrap merchants used to buy and sell valuable used commodities from homeowners. During World War II, Americans gladly recycled their tinplated steel cans for use in the war effort. Today, many people take equivalent items to a recycling or buy-back center, or leave them at the curb for reuse.

Each civilization has coped with the problem of waste in its own unique way. Today, refuse impinges on our lives, lifestyle, natural resources, and the environment with increasing intensity and in ways never previously foreseen.

Understanding current solid waste—its components and sources and how the components tend to be handled—lays the foundation for responsible action across the United States.

Refuse can be divided into two major categories: solid waste (SW) and hazardous waste. The former is usually not harmful to human health (exceptions are noted below). Typically, solid waste is, indeed, solid; it emanates from residential, industrial, commercial, mining, agricultural, and community activities. It consists of packaging, leather, rubber, textiles, lumber, metals, glass, paper products, yard wastes, and food wastes. Once used, packaging becomes part of solid waste.

In this chapter we will discuss the characteristics of solid waste with particular emphasis on packaging's role in creating paper, glass, plastic, and metal wastes. Our focus is on how each material is handled currently, according to the EPA's four waste-handling categories: source reduction, recycling, incineration, and landfilling.

The term "recycling" covers a variety of practices.

Industrial recycling, for example, may require that employees at a plant, office, or retail outlet separate and bundle corrugated and other paper products. These products are picked up by haulers or taken to a company warehouse for consolidation and marketing. Independently, haulers also sort through selected loads having a larger percentage of corrugated or other valuable products.

Residential recycling usually begins with the homeowner "source separating," that is, sorting trash by categories, such as newspapers and cans. The segregated material then is taken to voluntary drop-off or buy-back centers, where the material is placed in bins. Buy-back centers are manned and pay consumers for the materials.

Another approach to recycling is curbside collection. Millions of homes nationwide now have curbside recycling, run by municipalities, private haulers, and not-for-profit groups. With this system, collectors gather recyclables from bins at the residences on a regular basis. Several states, including New Jersey, Rhode Island, and Connecticut, require source separation and curbside recycling, as do numerous cities and counties in the United States. These programs focus on single-family residences. High-rise apartments have less storage space for recyclables and are harder to serve with curbside recycling.

GENERAL STRUCTURE OF SOLID WASTE

An estimated 160 million tons of municipal solid waste (MSW) were disposed of in the United States in 1987.[1] This waste emanated from residences (both single and multifamily); institutions (public and semipublic, including schools, prisons, and hospitals); commercial establishments (such as stores, offices, restaurants, hotels, and airports); municipal works (including street cleaning and tree trimming); and industrial refuse (particularly corrugated boxes and building and cafeteria wastes, such as towels). It does not include industry process waste. The components of MSW are shown in Exhibit 4.1.

It should be noted that, at times, other wastes are handled with MSW. These include:

- Municipal sludge
- Municipal waste combustion ash
- Industrial waste
- Construction, demolition debris
- Household hazardous waste
- Small quantities of industrial hazardous waste, such as from auto repair garages and dry cleaners.

Various sources anticipate that solid waste will continue growing at a rate of 2 to 3 percent per year. This is due to expected increases in population and consumption, augmented by greater use of plastic packaging (the most difficult to recycle currently). While there will, indeed, be some SW growth, it is anticipated that such waste handling methods as

Exhibit 4.1. Municipal solid waste composition.

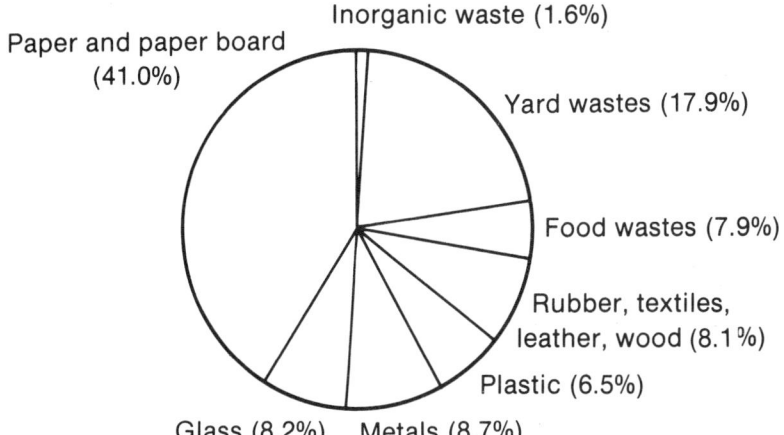

Source: EPA.

recycling, source reduction, and incineration will also increase, thus partially offsetting and delaying the need to send waste to landfills.

The general trends can be seen in Exhibit 4.2, which charts growth in key MSW components from 1960 on. We can make a number of observations, based on these statistics and other industry data:

- Glass has been declining as a percent of MSW.
- Plastics are expected to increase because of their convenience and potential for reducing the need for space in landfills.
- More yard waste will be composted by mandate so that it will decrease in volume and percentage terms.
- Tires, a significant problem currently, will be shredded and burned for energy recovery or made into other products.
- Office paper is increasing due to greater computer and high-speed copier usage. A higher economic value will make office paper more recyclable.
- Although not readily apparent from the above, packaging has declined as a percentage of total MSW by weight.

Exhibit 4.2. Gross discards* of material in MSW.
(In millions of tons.)

Materials	1960	1970	1986	2000
Paper and paperboard	29.8	43.9	64.7	86.5
Glass	6.5	12.7	12.9	13.4
Metals	10.5	13.7	13.7	15.9
Plastics	0.4	3.0	10.3	15.7
Rubber, Leather, Textiles, Wood	6.8	9.3	12.6	13.4
Food Wastes	12.2	12.8	12.5	12.3
Yard Wastes	20.0	23.2	28.3	32.0
Miscellaneous Inorganics	1.3	1.9	2.7	3.3
Totals	87.5	120.4	157.7	192.6

*Before materials recovery for recycling.
Source: Franklin Associates, Ltd.

Estimates put the gross per capita U.S. waste generation rate at 3.6 pounds each day in 1986. The residential rate was 2.5 pounds/person/day. The U.S. commercial rate ranged from one-half to one pound/person/day when recycling and energy recovery efforts were factored in.

In a populous area like New York State, the equivalent 1985 total MSW rate was 5 pounds/person/day. Other industrialized countries have waste-generation rates somewhat lower than that of the United States. For instance, in Japan it was 2 pounds/person/day in 1983, and in West Germany and Switzerland it was 2.6 and 2.4 pounds/person/day, respectively, in 1986; in Norway it was 1.7 and in Sweden 2.5 pounds/person/day.[2]

By the year 2000, net waste discards in the U.S. are expected to decrease slightly from 3.0 to 2.8 pounds per person per day. On the other hand, gross discards are forecasted to go up from 3.6 to 3.9 pounds. The difference will be handled by more recycling and burning. Exhibit 4.3 illustrates this pattern.

There are higher waste generation rates in more populated areas. They are accounted for mostly by larger newspapers and commercial refuse generated by more and bigger stores, restaurants, and factories.

Exhibit 4.3. Gross discards, recovery, and net discards of municipal solid waste.
(In pounds per person per day)

Products	1960	1970	1986	2000
Gross Discards	2.65	3.22	3.58	3.94
Materials Recycling	0.18	0.21	0.39	0.49
Energy Recovery	—	0.01	0.22	0.65
Net Discards	2.48	3.00	2.98	2.80

Source: Franklin Associates, Ltd., 1988.

PACKAGING SOLID WASTE

Packaging materials account for 33.8 percent by weight of municipal solid waste—the largest single category. It is comprised mostly of glass, paper and paperboard, metal, and plastic. Exhibit 4.4 shows the breakdown.

Paper and paperboard make up over 50 percent of packaging materials discarded. The use of paper packaging is expected to increase both in percentage and absolute terms. Plastic is expected to increase the most of any material, to 9 percent of the total from a mere 0.2 percent in 1960. Aluminum will go up slightly, as will steel. Glass is the only material expected to decrease as more of its applications are taken over by plastic.

Multilayer plastics provide more convenience (as with the squeezable ketchup bottle) and lighter weight, saving transportation costs. Plastic soft drink containers may also make inroads against aluminum as the price of the metal increases. Previously, plastic had essentially replaced glass usage in large soft drink bottles, and plastic and paperboard containers have supplanted glass milk bottles. Currently, there is an upswing in the use of plastic grocery bags, with a corresponding decrease in the volume of paper bag usage.

The data on plastic and other materials can be somewhat misleading if viewed only in terms of tonnage. One ton of plastic will generate more containers than a ton of glass or steel. Conversely, the decrease in glass tonnage implies an even greater market share loss in containers.

Exhibit 4.4. Gross discards of MSW products, 1986.

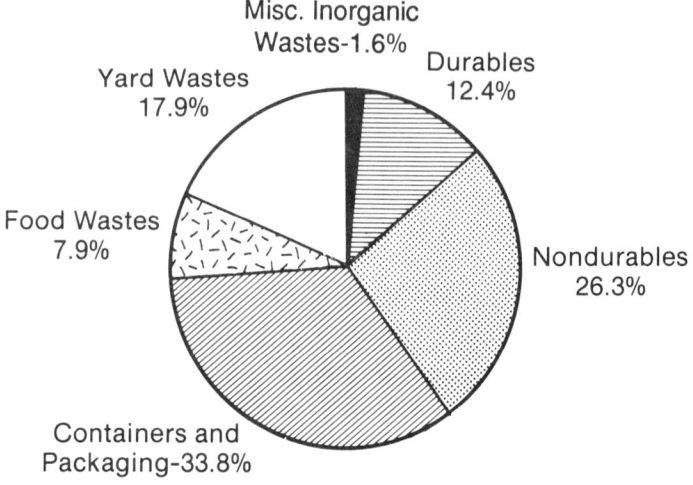

Source: EPA.

Steel will make some strides as a material for beverage containers, again because of aluminum's cost, but it will not regain its once dominant position. Three other factors will influence the increase in steel container usage: (1) The steel industry is beginning to provide strong financial support for steel can recycling. (2) Improvements in technology permit thinner walls in can construction and, therefore, the manufacture of lighter-weight cans. (3) Steel also is beginning to be used in combination with aluminum to create bi-metal containers. In some areas of the country, particularly steel-producing states like Indiana, all steel or bimetal cans are receiving strong consumer support.

One aspect of packaging vis-a-vis solid waste is its short life span. By its very nature, modern packaging is not needed for a long time. However, materials used in packaging last for an extended time. Under most circumstances, packaging endures well beyond the period and purpose for which it was designed. Nature's packaging, such as corn husks and banana peels, degrades at about the same rate as the product it contains. More durable packaging, such as returnable beer bottles (in Japan they get up to twenty return trips[3]), and consumer durable goods, such as refrigerators and cars, have a lesser or delayed impact on SW. Due to their weight, however, transportation costs for these items are higher. These and other factors have a bearing on solutions to packaging solid waste.

PRIMARY SOLID WASTE COMPONENTS AND HANDLING

From residences, schools, businesses, and industrial firms come discarded paper products: sacks, bags, shoe boxes, corrugated containers, food packages, newspapers and the like.

Paper and Paperboard Waste

At 36 percent (after recovery and recycling) of MSW in 1986, paper (including paperboard) is the largest single component of the municipal waste stream. Continued growth is expected. Sixty-three percent of paper discards goes to landfills or is incinerated, as shown in Exhibit 4.5.

Almost one-half (48 percent) of all paper produced is employed in packaging applications—6 percent in paper packaging and industrial converting, 42 percent in corrugated containers and other boxboard products.

Exhibit 4.5. Paper in the municipal solid waste stream.

Year	Paper Consumed (Mil. Tons)	Paper in MSW (Mil. Tons)	Percent of MSW	Percent Disposed of*
1960	38.5	24.5	30.0	63.6
1970	56.0	36.5	32.4	65.2
1986	79.8	50.1	35.6	62.6
2000 (est.)	106.5	66.0	39.1	62.0

*That is, sent to landfill or incinerator (not recycled).
Source: EPA: *Appendices*, Table A.A-1, p. A.A-2.

Source Reduction

Greater efforts to reduce paper use would have a significant impact because paper currently represents over 41 percent of the waste stream (before recovery and recycling). Reduced usage would be especially useful in urban areas, where people consume more per capita and, concurrently, where the pressure on landfills is the greatest.

However, the trend appears to be in the opposite direction. In recent years, paper usage has increased with the growth in popularity of convenience foods and more individualized, smaller frozen-food and other packages.

Recycling

Recycling accounts for 28 percent (28.6 percent in 1988) of all paper discards. Of all paper packaging, corrugated containerboard—40 percent of it—is most often recycled. Exhibit 4.6 shows the recovery rates for all paper categories. The recycling success rate for paper is due to a material of relatively uniform quality and large, segregated quantities. These factors decrease sorting and labor costs. The basic recycled corrugated product is comparatively clean and has

Exhibit 4.6 Percentage of paper recovered by category.

Paper Category	Total Recovery as % of Consumption (1986)	Postconsumer Recovery as % of Gross Discards (1986)
Paper Packaging		
Industrial Converting	10.1	5.7
Containerboard (corrugated)	47.0	41.1
Boxboard, other Paperboard	20.6	5.6
Total Paper	27.9%	22.6%

Source: EPA: *Appendices*, Table A.A-3, p. A.A-7.

a long fiber content, which are important considerations for future product applications.

The end result is an established infrastructure of scrap dealers and entrepreneurs that has existed for decades to deliver material to paper mills at a profit. Similar networks are being created for computer and copier paper because of their high value and other similar characteristics.

Established markets and uses for recycled packaging exist already. Old corrugated containerboard is frequently employed to create recycled paperboard. Examples include boxes for cereal and dry macaroni. Used corrugated also goes into linerboard and the corrugating medium for new containers.

Three other sources for creating recycled paperboard are old newsprint, mixed paper (unsorted, mixed paper), and mixed pulp substitutes (such as trimmings from the conversion of paper into envelopes). Old newspapers can also be made into molded pulp. Exhibit 4.7 summarizes the uses for recycled paper in 1986.

Technically all packaging grades of paper can be made with at least some recycled fiber content, but there are specific problems involved in using recycled paper. What recycled paper can be regenerated into depends on the quality and type of the recycled product and on the requirements of the new one.

The amount of contamination (food, foil, plastic, ink) and the strength of the fiber are two key considerations. Generally, recycled paperboard is more tolerant of contamination than other paper products.

Additional barriers to using recycled material include higher unit costs due to lower volume and consumers' fears that recycled products are inferior. A recycled fiber cereal carton, gray on the inside, is acceptable to consumers. It might be harder to sell them grayish-colored ice cream cartons, even if they were coated.

A key problem in handling recycled paper may be the paper mill itself. Those mills that can handle recycled materials were operating at 97 percent capacity in 1988.[4] But paper mills are not always located near a major source of recycled fiber supply, i.e., a large metropolitan area. And

Exhibit 4.7. Composition of recovered paper in domestic paper and paperboard mills, 1986. (In thousand short tons.)

Product Category	Mixed Papers	News-papers	Corrugated	Pulp Substitutes	High-Grade Deinking	Total
PAPER						
Newsprint	—	1,364	—	1	—	1,365
Printing & Writing	—	—	4	910	342	1,257
Packaging & Industrial Converting	25	10	107	183	5	331
Tissue	97	191	197	655	1,049	2,189
Total Paper	122	1,566	308	1,749	1,396	5,142
PAPERBOARD						
Kraft, Bleached & Unbleached	—	—	1,561	112	10	1,684
Semichemical	50	—	1,607	48	—	1,704
Recycled	1,525	1,293	4,931	796	162	8,706
Total	1,575	1,293	8,099	956	172	12,094
CONSTRUCTION PAPER & BOARD	348	259	277	57	1	892
TOTAL PAPER INDUSTRY	2,045	3,118	8,634	2,761	1,570	18,128

Source: American Paper Industries, Inc. *Paper, Paperboard, Wood Pulp Capacity,* 1986–1989.

although a new virgin newsprint mill can cost $400 to $500 million and take five years to build, paper companies have not felt that the supply of recycled fiber was consistent or large enough to justify retrofitting an existing virgin mill with special equipment to handle recovered paper at a cost of $20 to $30 million.[5] Ten of the approximately twelve new newsprint mills currently planned for construction are based on the use of virgin stock. (Some may be altered during construction to take recycled fiber.)

The presence of ink also hinders recycling. In some instances it may be difficult to remove from the paper, as with impressions made by laser copiers, which bond the ink to the paper by high heat. Other pigmented inks made with toxic heavy metals such as cadmium create toxic sludges. These must be treated as hazardous waste and sent to specially designed and permitted landfills.

Another current difficulty with paper recycling is that supply exceeds demand. This has aided exports, which claim an increasingly greater share of recovered paper. In 1970, recovered paper was 2.8 percent of net paper export; in 1987, that figure rose to 22.5 percent.[6] For Japan, Mexico, and other timber-poor countries, used paper is an inexpensive substitute for virgin stock.

The oversupply of recovered paper and consequent lower market prices not only augment exports but also put pressure on collectors. Traditionally, paper has been the largest volume item collected. It is not the most profitable product to recycle, however. But, paper has made other recycling programs viable because consumers recycle higher margin cans and bottles along with paper.

Paper exports in 1989 are expected to reach 7 million tons. Historically, they have been much lower, as Exhibit 4.8 shows. For comparative purposes, paper recovery rates for other countries are as follows: France—14 percent; West Germany—40 percent; United Kingdom—29 percent; Sweden—40 percent; Japan—50 percent.[7]

Incineration, Energy Recovery

Paper is a good candidate for energy recovery through incineration because it has a high heating value (in Btu/lb.) and low moisture content. Also, little ash remains after

Exhibit 4.8. Paper exported for recycling.

Year	Net Export Tons	Net Exports As % of Recovered Paper
1960	96	1.1
1970	341	2.8
1980	2,577	17.1
1986	3,650	20.1
1987	4,295	22.5

Source: EPA: *Appendices,* Table A.A-5, p. A.A.-14.

burning (uncoated paper 10 percent ash, coated paper 30 percent). However, reluctance to use the burning method is increasing as recycling becomes more entrenched. The use of heavy metal in inks also creates concern about the emissions and ash resulting from incineration.

Incineration could be a good solution to the periodic paper gluts. An additional advantage is the volume reduction of paper to ash (a 9 to 1 ratio).

Landfilling

Paper does not decompose rapidly in today's landfills, which minimize moisture content. Additionally, paper takes up 38 percent more volume than it does weight, for a total of almost 48 percent of landfill volume.

Without adding other costly steps such as shredding, paper cannot be compacted more than it is. Thus, its impact on the landfill is likely to remain significant unless it is sidetracked before getting there.

Plastic and glass have worse ratios of weight to volume, as Exhibit 4.9 shows. But because they do not take up as much of current landfill space, their volume impact is not as severe as is that of paper.

Glass Wastes

At 8.2 percent (in 1986), glass is the third largest portion of MSW. It was 11 percent in the early 1980s and is predicted

Exhibit 4.9. Weight-to-volume ratios: paper, glass, and plastic.

Material	Percent of Landfill	
	By Weight	*By Volume*
All paper	34.5	47.5
Plastic	4.8	16.3
Glass	2.4	.9

Source: EPA: *Appendices*, p. A.A.-48.

to be 7 percent by the year 2000. Lighter weight aluminum and plastic products are causing this shift. Their impact on glass is significant in that beverage and food containers account for 70 percent of all glass produced and for 90 percent of the glass in MSW. This compares to the 48 percent of paper products that goes into packaging applications. Exhibit 4.10 shows the decline in glass as a contributor to MSW.

Households provide the main source of glass waste. All colors of container glass—clear, brown, green—are now recycled. Of the 10.7 million tons discarded in 1986, about 1 million are crushed into cullet and used as feed for generating new glass. This is more than 40 times the 1970 level. As

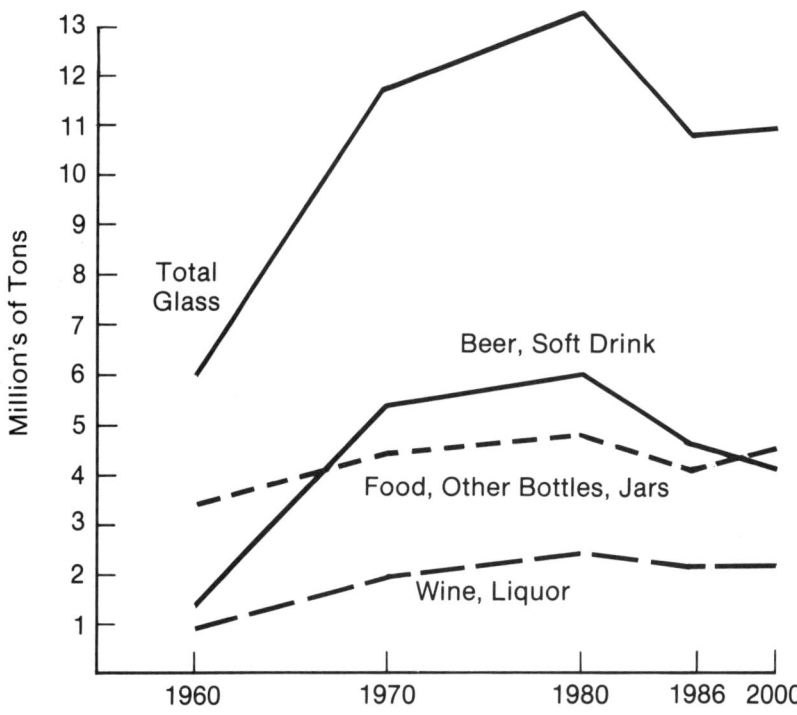

Exhibit 4.10. Glass containers, packaging in MSW, 1966—2000.

Source: Franklin Associates, Ltd.

of 1986, the U.S. glass industry was employing about 74 percent of its 13 million-ton cullet capacity. Existing technology exists to permit 100 percent cullet use in glassmaking furnaces. The current trend, however, is more toward 40 percent. One key to higher cullet use remains a consistent supply.

Recycling

The benefits of using recycled cullet stock are several. The energy used is decreased by 15 to 32 percent, mining wastes by 80 percent. Water usage is cut in half.

The quality of glass sorting significantly affects its market price and end use. Mixed colors are usually detrimental to remelting efforts from an appearance and application standpoint. Cullet must also be relatively free of metallics, plate and other non-container glass, as well as other contaminants. Some equipment exists nationally to grind glass, remove metal, and clean cullet. But machinery will not remove all non-metallics nor color-sort cullet economically.

The biggest market for recycled glass cullet is new containers. Currently, about 25 percent[8] of glass-container raw material comes from cullet. Because any potentially harmful bacteria are killed by the 2,700 degree Farenheit temperatures, old beverage bottles and food containers can be made into new ones, again assuming colors have been adequately separated and contaminants are low enough.

Other markets include construction materials. As an addition to bricks, cullet permits lower temperatures and shorter firing time, thus decreasing costs. It can be used in floor and wall tiles. It can also be combined with asphalt for roadbeds and provide glass beads for highway marking. The roadbed application, however, calls for a tremendous supply and provides little added value to the product. Fiberglass insulation, a 1.5 million ton[9] market annually for cullet, is another use for recycled stock. Fiberglass is now produced with 10 to 15 percent glass cullet.

Current glass disposal methods include source reduction, recycling, and landfilling. Incineration is not an option since glass is not combustible.

Some reduction is practiced rigorously in Norway and Denmark, where only a limited number of returnable beverage container configurations are allowed. This facilitates reuse by a variety of bottlers.

In the United States, convenience and competition have led to a wide variety of container types, shapes, and colors. Most are made from very similar glass chemistries, a fact that aids recycling. As of 1986, only 15 percent of soft drinks (measured in volume) reached consumers in refillable bottles. For beer, the volume was only 6 percent.[10] In Japan, 66 percent of all bottles are used three times, on average, with beer bottles being used up to 20 times.

Recycling claims 27 percent of glass in the U.S.[11] For comparison, Exhibit 4.11 shows recovery rates for other countries. Some say that 50 to 90 percent of glass containers could be recycled—theoretically, 100 percent. Rhode Island has already achieved a 60 percent level.[12] The economics of alternative disposal methods and mill specifications for recycling, however, would suggest that somewhat less than 90 percent is a more practical goal.

Several glass recycling methods are in common use. Some are similar to those for paper. Most glass programs run concurrently with programs for other materials such as newspapers and cans.

Exhibit 4.11. Glass recovery rates.

Country	Glass Recovery Rate*
Netherlands	53
Italy	25
West Germany	39
Japan	17 (1984)
France	26
United Kingdom	12
Switzerland	44 (all glass)

*1985 data used unless otherwise noted. Data apply to nonrefillables only. In Europe, most glass containers are refillable.

Source: EPA: *Appendices,* Table A.B-6, p. A.B-25.

There are many voluntary drop-off centers and bottle banks to which residents take glass containers, which are generally separated and put into different bins by color: amber (brown), flint (clear), and green. Some communities require glass to be separated from other refuse for purposes of curbside collection.

Glass Processing Facilities

Once glass is collected, the 27 percent that is recycled rather than landfilled goes to processing facilities. These are owned and operated by the glass producers, by independent operators, or by partnerships of producers and processing facility operators. Such centers are spreading around the country.

They all have a similar function: to prepare glass for reintroduction to the glass companies' furnaces. With varying degrees of automation, these facilities (generally costing between $300,000 and $500,000)[13] remove labels, paper, plastic, metal, and other contaminants. Typically the centers specialize in one type of glass, separated by color prior to their receiving it.

The bottles are then crushed, granulated, or ground into powder and sent through a final screening to remove remaining contaminants. This cullet is then ready for the furnace.

The early glass processors by and large were glass producers or intermediate glass processors (IGPs). The latter tended to focus on preconsumer glass scrap from brewers, bottlers, dairies, and makers of flat glass. As such, they were usually located in more remote areas, near their suppliers. The newer IGPs, by contrast, are seeking postconsumer material. In some instances, the glass processing becomes part of a broad-based materials recovery facility (MRF). In either case, they situate in or near urban areas, being dependent on a consistent quality and quantity of material. Because current cullet prices are competitive with prices for virgin material, more IGPs and MRFs will likely increase the percentage of cullet as glass container raw material significantly.

Markets

Having adequately sized markets for recycled glass does not appear to be the problem it is for paper. Cullet is not expected to be in oversupply until national recycling rates reach 50 percent or more. This should provide ample time to strengthen the above and other markets.

PLASTIC DISCARDS

Plastics constitute about 7 percent of MSW by weight; it is estimated that they will increase to 9 percent by the year 2000. Plastic packaging used almost one-third of the plastic consumed in the U.S. in 1985, or about 13 billion pounds. Behind paper and metal, plastic is the third-largest segment of packaging (13 percent). The major components are shown in Exhibit 4.12.

The six major plastic resins accounting for 96 percent of packaging applications and types are shown in Exhibit 4.13.

A wide variety of plastics is available for an increasingly diverse and expanding set of applications. Through chemical modifications and enhancements, resins can competitively fulfill packaging requirements in a manner not possible previously. The resulting features include: lower cost; lighter weight, enabling, for example, larger soft drink bottles; and improved barrier properties providing more protection against odor, flavor, air, and solvent contamination.

Sources of plastic packaging discards are residential, industrial, and commercial, as with the other materials. Residential refuse predominates since much packaging is for consumer products. Sizable quantities, however, are generated by industrial users. Examples of these include office buildings with cafeteria operators and assembly plants with bulk shipments of products in plastic stretch-film overwrap. Commercial establishments such as restaurants, auto repair facilities, hospitals, and prisons also create significant discards.

Even though plastics are only 7 percent of the waste stream by weight (16 percent or more by volume), they are a

Exhibit 4.12. Major components of plastic packaging.

Packaging Type	Applications	Percent of Total Plastic Packaging
Films	*Trash, grocery bags *Boil-in-bag pouches *Refrigerator overwraps	35
Bottles	*Variety of foods, beverages, edible oils *Motor oil, detergent, bleach, other products *Transparent, nontransparent bottles	27
Other Rigid Containers	*"Tubs" for foods like yogurt *Polystyrene foam-egg cartons, insulated cups, fast food "clamshells"	24
Coatings, Closures	*"Barrier" coatings to protect container from moisture, oils, chemicals *Screw tops, snap-on cup lids	14
Total		100%

Source: *Modern Plastics*, January 1987, pp. 56–62.

very visible target of consumer groups and legislators. One reason for this is rapid growth in their use (from 4.7 billion pounds in 1970 to 12.8 billion total pounds in 1985)[14] and their visibility as litter. Another is the lack of a comprehensive infrastructure for their recovery and reuse. Of the 12.8 billion pounds used in packaging, only .2 billion pounds (1.6 percent) were recovered for recycling, most of this being represented by PET soft drink containers and HDPE milk bottles. (In 1985, 20 percent of all PET soft drink containers

Exhibit 4.13. Packaging resins and their applications.

Packaging Resin	Applications	Percent
Low Density Polyethylene (LDPE)	Films, Bottles	33
High Density Polyethylene (HDPE)	Films, Bottles	31
Polystyrene (PS)	Other Rigid Containers, Closures	11
Polypropylene (PP)	Other Rigid Containers	9
Polyethylene Terephthalate (PET)	Bottles, Boil Bags	7
Polyvinyl Chloride (PVC)	Bottles, Films	5
Other		4
Total		100%

Source: *Modern Plastics,* January 1987, pp. 56–62.

were recycled.)[15] The remaining 11 billion pounds were landfilled. Packaging represents one-half of all plastic waste.

Source reduction of plastics receives attention from packaging firms because of their desire to lower costs. The major emphasis is on reducing volume and toxicity.

Recycling

Recycling of some industrial plastic scrap has been going on for years. For example, most synthetic carpets and carpet backing are made from *PET* waste. Postconsumer plastic recycling has been inhibited by its low density in relation to weight, making collection difficult and expensive. Other contributing factors include contamination (from food, etc.), the lack of large collection systems and established markets, and the difficulty of segregating postconsumer plastic waste (everything from cheese wrappers and yogurt cups to soft drink and milk containers, including single and multilayer materials). It is ironic that just as aseptic and other sophisticated packages are extending food shelf life and increasing convenience, they are coming under fire for being so durable after serving the consumer.

Adding to collection expenses is the requirement to sort by plastic type. Products can be fabricated from mixed plastics (discussed below). But, higher value-added end uses are possible when the material is segregated. An industry-designed container coding system has been introduced to assist the sorting process.

Other, more mechanized means of separating plastics and removing contaminants also are being developed. These use flotation and electrostatic systems.

Significant alliances and investments are now being made to enhance the recyclability of plastics and to address some of the difficulties mentioned above.

Typical liaisons involve state governments and major chemical firms, as for example, that between the Illinois Department of Transportation and the DuPont Company. Together, they are seeking to create road barrier and other transportation-related markets for recycled material. Other combinations include major waste management firms and chemical or plastics processing companies. In addition, chemical and plastics producing firms are forming consortia to design and build recycling plants. Trade associations and university-firm relationships have also been formed to conduct research, particularly on unsorted plastics. Under trial in institutional cafeterias and fast food restaurants are experiments in which consumers separate the plastics when disposing of their trays. These plastics are then reprocessed into pellets for manufacture into a variety of molded products.

Markets for recycled plastics do exist in significant numbers. Other than carpeting, they include nonfood containers (a new area), package strapping, high temperature PET foam, woven geotextiles for roadbeds and erosion control, roofing membrane, plastic lumber, fencing, pipes, yard furniture and stadium seating, and base cups for soft drink bottles. More sophisticated applications, such as engineered plastics for automotive parts, are also possible.

Incineration, Energy Recovery

Plastics can release as much as 15,000 to 20,000 Btu/lb.—or up to four times the average energy potential of MSW, twice that of Wyoming Coal.[11] This high heat value is a valuable resource in waste-to-energy plants. In older incinerators, not designed for the richer fuel, the hotter burning wastes can cause operational difficulties. There is concern about potential adverse effects from burning plastics, or any material, in improperly operated incinerators.

Landfilling

Considering that 11 billion pounds of plastic packaging were landfilled in 1986, there is certainly potential for reducing that amount. It can be said, however, that fully compacted plastics take up less landfill space than equivalent rigid bottles. Studies have found that the majority of glass bottles in a landfill are not compacted.

ALUMINUM AND STEEL

Aluminum can recovery is a real MSW success story. With a 55 percent recycle rate nationwide in 1988, aluminum cans have achieved the highest recycling level of any packaging material. In terms of tonnage, aluminum cans represent about 19 percent of all metal packaging material discards. See Exhibit 4.14.

Various estimates indicate that aluminum cans will increase to one million tons by the year 2000. Steel cans will decline to 1.8 million tons by 2000, significantly below the 1986 level of 2.4 million tons. However, the steel industry is increasing its share of the container market by promoting bi-metal cans (aluminum ends with steel can bodies) and by its strong financial support of recycling.

As with paper and glass, metal for recycling comes from residential, commercial, and industrial sources. As with glass, residences are the largest source, with beverage and food cans predominating.

Exhibit 4.14. Packaging materials discards.

Material	Tons (mil. '86)	Percent to Total Metal	Percent to Total Metal Waste
Steel—beer, soft drink cans	.1	3	.1
Steel—food, other cans	2.4	65	1.7
Steel—other packaging	.2	5	.1
Aluminum—beer, soft drink cans	.7	19	.5
Aluminum—foil, closures	.3	8	.2
Total	3.7	100	2.6

Source: EPA: *Appendices*, Table A-G-2, p. A.G. 2.

Various grades of aluminum and steel are used in packaging. Aluminum-container stock consists of four separate types. Steel is coated with tin or plastic materials for corrosion resistance.

Impurities in the metals themselves, or from mixing during the collection process, create difficulties in recycling and reuse. De-tinning, or removing the tin coatings from used cans, is an economically viable operation, as is magnetic separation of steel from aluminum. Mixing of different grades, however, causes problems in generating new metal chemistries that sometimes depend on small differences in the various elements for property enhancements. Bi-metal cans could be difficult to sort out from aluminum or steel varieties. For these reasons, mills rely on in-house scrap or on known scrap vendors, who control the quality of refuse sold to mills.

Exports of used aluminum cans have skyrocketed recently, up 44 percent in 1988 over the prior year, according to the U.S. Department of Commerce.[17] This trend is understandable, given the large U.S. market, the unfavorable balance of trade, and the energy and bauxite savings that accrue to resource-poor countries.

Steel cans account for 4 percent of the approximately 80

billion beverage cans produced in 1987. Tin-plated steel cans have an 8 percent share of the soft drink container market and less than 1 percent of the beer container market.[18] Currently, only 3 to 5 percent of our steel beverage containers are being recycled. The Steel Can Recycling Institute, supported by many U.S. tinplate producers, was formed in 1988 to augment recycling. Factors that will help the industry include relatively stable pricing in recent years, steel's magnetic properties, and its cost advantage over aluminum.

Source Reduction

The creation of thinner materials that do a comparable job has been the most significant recent source reduction in metal packaging. For example, in 1985, 26.6 aluminum cans could be fabricated per pound. In 1988 that number increased to 28.3 cans per pound—a 6.4 percent increase in yield in four years.[19] Steel cans are also decreasing in wall thickness. As a result, less raw material is employed and less petroleum is used to transport them throughout their life cycle. Tighter quality control procedures and material specifications are needed, however, because more is being demanded of the metal.

Recycling

Typically, the recycling of metal containers and other postconsumer metal packaging is conducted in similar fashion to paper and glass recycling. The drop-off, buy-back, and curbside recycling mechanisms often exist in tandem with collection for other material.

Aluminum's historically high value provides ample margin for collection costs. Aluminum returns can generate enough profits to make entire recycling efforts viable, since glass and paper collection work at close to a break-even position.

Producers have strong economic incentives for recycling. Charging the furnace with recycled material saves about 95 percent of the energy and about 4 pounds of bauxite for every pound of used containers. For these reasons, the aluminum companies have been avid supporters of recycling for twenty years. Exhibit 4.15 illustrates the recycling rates of this industry.

The aluminum industry's unique contribution to recycling has been company-sponsored buy-back centers, often placed in strategic urban locations such as shopping centers. Operated by aluminum company employees, the centers reimburse consumers who bring in cans. Ordinarily only aluminum is accepted at these centers. Because of these programs and aluminum's intrinsic value, a whole subculture of collectors has arisen, picking up used cans everywhere,

Exhibit 4.15. U.S. Aluminum Can Recycling Rates.

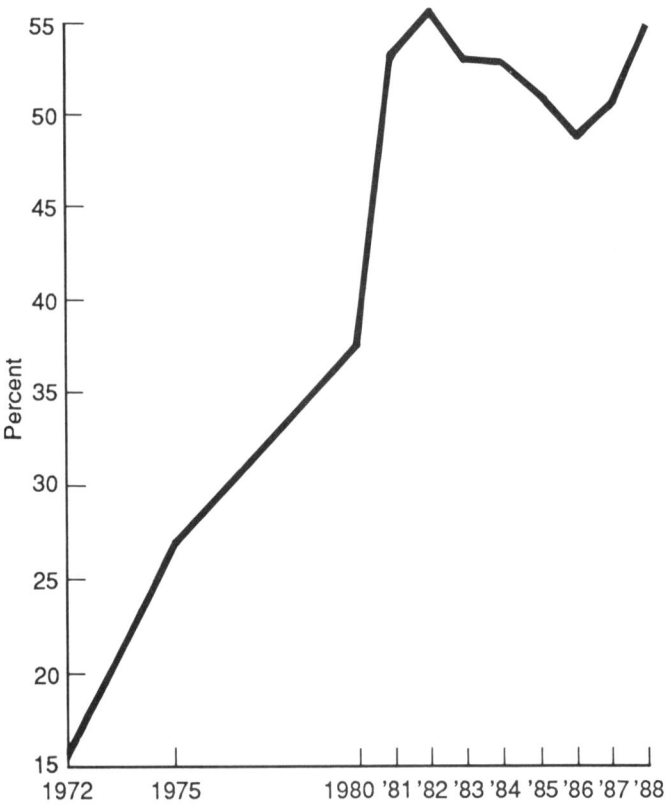

Source: *Phoenix Quarterly*, Spring 1989, p. 9; *Aluminum Recycling Facts* (The Aluminum Association, Inc.)

and thereby lessening beverage can litter. There are also so-called "reverse" vending machines that accept empty beverage cans and dispense money in return.

Steel of all types has been recycled for a long time. In modern society, recycling embraces used autos and other more durable goods. The scrap yards and dealers have created a sizable infrastructure for recycling steel, one that includes over 6,000 U.S. scrap and waste materials businesses,[20] in addition to smelters and steel mills. This same network makes use of other materials as well. In fact, as a commodity steel is the most recycled of all materials. Of course, others employ this same network. Scrap collectors may sort and otherwise process material or sell it to another business for processing. The sorted metals are then used by foundries, smelters, and basic steel or "mini" mills to create new products.

These products, vis-a-vis packaging, can be new food and beverage containers or other packaging materials. New cans are, of course, their most usual reuse. Being thin, foil is volatilized or burned up in remelting. Because of high prices and manganese and magnesium content in used aluminum can stock, it is not often used for casting into new aluminum alloy products.

The aluminum can industry is expected to grow from 81 billion cans in 1988 to 120 billion in 1995.[21] Thus, the end-use markets for recycled material will be ample. In fact, the aluminum industry's goal for recycling is 75 percent by 1995.[22] Perhaps it could be even higher.

Incineration

Metals are low in energy value. Further, burning some of them creates potential for detrimental air emissions and residual ash that is hazardous. These result from metallic compounds or inks and paints covering them. Metals with a lower melting point relative to others in incineration will also create slag in a mass burn or other types of furnaces and incinerators. This slag causes premature breakdown of furnace linings. Thus, metal that is not separated or sorted out of refuse prior to burning does not contribute to economic operation or sound waste management practices.

Landfills

Although aluminum has good recycling rates, about 45 percent of this valuable resource still ends up in landfills. A variety of ways are used to sort out metals. The most common method is magnetic separation of ferrous materials. Air jets, hand/visual sorting, screens, and water flotation techniques are also employed. Metals not culled out of material that is subsequently shredded for size reduction do damage to the shredding equipment.

At landfills hand sorting does take place to remove significant amounts of metal because of its economic value, its established infrastructure for handling, and existing markets for reclaimed material. More of this sorting appears to take place at smaller local landfills that handle lower volumes, and where employees have more time available for such separation. The presence of lead, cadmium, and other "heavy" metals in landfills can lead to the creation of hazardous wastes when there is contact with battery acid.

Market Incentives

The metal scrap market is subject to volatility, as are the paper, glass, and plastics recovery industries. Similarly, lower prices do not lead to higher demand in contradiction to usual economic theory.

The major incentives for metal recycling are economic. With markets already in place, a healthy economy supports the scrap industry infrastructure. The aluminum industry in particular has an adequate profit motive to foster even greater levels of recovery.

REFERENCES

A significant amount of data for this chapter has been obtained from the Environmental Protection Agency (EPA), *The Solid Waste Dilemma: An Agenda for Action*, September

1988. Its two volumes are entitled, *Background Document* and *Appendices A, B, C.* They are referred to in the notes as EPA: *Background* and EPA: *Appendices*. The other major reference cited is Franklin Associates, Ltd., *Characterization of Municipal Solid Waste in the United States, 1960 to 2000 (Update 1988),* U.S. EPA, Office of Solid Waste and Emergency Response, March 30, 1988. It will be referred to as *Franklin: Characterization.*

1. EPA: *Background,* Table 1–1, pp. 1–6.
2. Allen Hershkowitz, *Garbage Burning, Lessons from Europe: Consensus and Controversy in Four European States* (Inform, Inc., 1987), p.13.
3. Allen Hershkowitz and Eugene Salerni, *Garbage Management in Japan: Leading the Way.* (Inform, Inc., 1987), p. 45.
4. Ben Harvey, "Access to Export Markets Essential for Paper Recycling," *Recycling Times,* May 23, 1989, p. 6.
5. Ken Birgenheimer, "Newsprint Market Lags Supply But Users Are Responding," *WasteTech News,* April 24, 1989, p. 14.
6. EPA: *Appendices,* Table A.A–5, p. A.A–14; Harvey, *op. cit*
7. EPA: *Appendices,* p. A.A–41.
8. Steve Apotheker, "Glass Processing: The Link Between Collection and Manufacture," *Resource Recycling,* July 1989, p. 43.
9. *Ibid.,* p. 42.
10. Chaz Miller, Glass Packaging Institute, Washington, D.C. Telephone conversation, November 16, 1989.
11. Apotheker, p. 43.
12. Chaz Miller, November 16, 1989.
13. *Ibid.,* p. 40.
14. EPA: *Appendices,* p. A.C–413.
15. *Plastics Recycling: From Vision to Reality* (Plastics Recycling Foundation, Center for Plastics Research, 1968), p. 2.
16. EPA: *Appendices,* Table A.A–9, p. A.A–46.
17. "The Amazing All-Aluminum Can," *Phoenix Quarterly,* Vol. 21, No. 2 (Spring 1989), p. 10.

18. Jefferson Reiter, "The Can of Steel," *Beverage World* reprint, August 1988.
19. *Phoenix Quarterly*, Spring 1989, p.9.
20. EPA: *Appendices*, p. A.G–25.
21. *Phoenix Quarterly*, Spring 1989, pp. 10,11.
22. *Ibid.*, p.10.

5

Management Agenda

We must be committed to the goal of fostering, supporting, and participating in wise solutions to the solid waste crisis. To achieve this, we must pursue strategies that promote astute governmental policies related to solid waste at the same time that we reduce the load that our companies place on the solid waste system.

We urge American businesses to consider the following actions, as they move toward initiatives appropriate to their particular industries and business practices.

I. Work to reduce the impact of packaging on solid waste without compromising package effectiveness for the consumer.

Ordinarily, packaging cost-reduction work results in source reduction—that is, the use of less material in packaging. This is not enough in the present situation. Options for packaging changes must also be included: design for recycling, use of recycled materials, and design for benign incineration and landfill.

The following actions have little or no effect on the appearance or effectiveness of packages and should be implemented. They cost little, but contribute much to efforts to improve the solid waste situation.

- Reduce the thickness or weight of packaging to achieve source reduction.
- Apply codes to plastic bottles and cups to identify the type of plastic. This will facilitate recycling. (Laws requiring the codes have been passed in many states.)
- Eliminate the use of all pigments formulated with cadmium, lead, and chromium to decrease the toxicity of recycled products, solid waste, and incinerator ash.
- Increase the use of recycled materials in packaging.
- Work to use more recyclable materials, insofar as such materials do not defeat the goal of source reduction.

Package designs can be altered to foster recycling through conversion to more highly recycled classes of materials, to single-material packages, or to systems of materials that are easily separated for recycling.

In designing for recycling, two cautions must be observed: (1) the design changes must not result in increased weight or volume of packaging (which defeats source reduction); and (2) the design should not be dependent on a particular recycling technology. This applies especially to packages made from multiple materials. Multilayer packaging is almost always lighter and smaller in volume than an equivalent monolithic package. Technology for separating materials and for using recycled mixed materials is evolving rapidly. The recycling infrastructure in the United States is growing and changing quickly. Efforts to constrain packaging to current recycling structures can cause source increases and inhibit the development of a broader recycling technology.

> II. Introduce innovative packaging to increase product acceptance through environmentally superior packaging forms.

The present high profile of solid waste offers opportunities for restaging the form in which products are offered to consumers. Pouches can be substituted for bottles. Products can be offered in concentrated forms that require smaller packages. European fabric softeners offer one excellent example of environmentally based packaging. Pouches of concentrate for refilling plastic bottles have achieved market success. Market research on the viability of such products will be difficult, as with all highly emotional and visible areas. But the opportunity for building sales through astute design changes has been exploited in Europe, and such opportunities are likely to exist in the United States. Indeed, some consumer product firms are pursuing the possibilities.

> III. Work toward comprehensive management of all solid waste.

Source reduction and recycling of office paper, cafeteria wastes, and plant waste are important parts of responsible corporate action. *These efforts will not only reduce solid waste but also provide leadership in shifting public focus to the overall problems of solid waste—and away from a counterproductive narrow focus on packaging.*

> IV. Use more recycled materials in packaged product design.

Increased community recycling programs have flooded the market in materials for recycling. Creating markets for these materials both decreases product costs and helps the environment. Modification of specifications can facilitate this.

> V. Establish a broad base of information sharing and decision making within the company and industry.

For many companies, the most visible environmental problems relate to plant emissions, an internal operations issue. The solid waste problem is quite different, in that actions to alleviate it directly affect the form and packaging of the product delivered to customers. Thus, it is a more complex business issue, one that requires a broader base of decision making within the company. It is helpful to identify a group of individuals within the company whose responsibilities bear on solid waste, packaging, and government regulation. This would include personnel from the legal, corporate relations, purchasing, package engineering, production, and corporate quality departments. Representatives from these areas should be informed and consulted on actions contemplated.

These people will also work with colleagues within the industry on a national and international basis to implement the best available technology and systems.

VI. Implement effective consumer education.

Because popular perceptions of packaging and solid waste are often based on inaccurate information (see Chapter 2), we must act to prevent incorrect views from forcing us into inappropriate packaging options.

The issue is becoming urgent.

Advertising is a particularly important basis for consumer understanding. As marketers attempt to gain competitive advantage through environmentally benign packaging, there will be a temptation to cater to the superficially appealing but technically unsound misconceptions consumers often have about packaging and the environment. While this will offer short-term advantage, it will perpetuate unsound policies—and could backfire. As consumer understanding grows, misleading statements run the risk of casting the manufacturer in an opportunistic and irresponsible light.

Avenues for consumer education include advertising, donations of educational materials to schools, and participation in, and guidance of, public interest groups. This book contains information that you can use in your personal and public advocacy as a community leader to bring the appropriate concerns of the public to productive action.

VII. Foster development of the recycling infrastructure.

The role of governments—local, state, and federal—in collecting and disposing of solid waste and in passing laws affecting solid waste makes them principal players in fostering recycling. A recycling infrastructure requires the interdependence of public and private operations. Governmental incentives can encourage recycling by fostering collection and creating markets. These warrant more detailed understanding.

Collection incentives include:

- Tax credits, exemptions for equipment purchases
- Grants for recycling promotion and equipment
- Low-interest loans
- Industrial bonds
- Accelerated depreciation

These incentives, to be effective, should be based on local industry needs and other conditions. For instance, tax credits in a state with low taxes do not foster collection.

At the local, state, and federal levels, guidelines (sometimes mandatory) are being instituted for procurement of products with recycled content. The EPA provided leadership in this area by publishing purchasing guidelines in 1988. Some directives even permit paying a premium to support purchases of recycled materials.

The private sector, through individuals and firms, can encourage recycling markets by requesting and buying recycled paper products. The need to create markets, as a part of a successful infrastructure, has become more pressing as curbside recycling programs are flooding the market.

Finally, government bodies can use legislation to encourage more recycling, including municipal curbside programs. The starting point for municipal curbside programs is legislation or ordinances requiring curbside collection. This approach typically has goals for the level of participation over a specified time period. One or more bins are provided at either the municipality's, hauler's, or resident's expense. The refuse may be commingled (i.e. unseparated) for one bin or source-separated (by the resident) for several bins. If sorting by the resident is mandated, all colors of glass are usually put in one bin—for separating at a later stage.

VIII. Participate in government policy formation.

Solid waste disposal is a political crisis in densely populated states and a political issue of high visibility in all states. It will be an agenda item in at least 30 state legislatures in 1990. The solid waste crisis is caused by landfills becoming filled at a time when siting new low-cost local landfills is very difficult. This is leading to dramatically increased disposal costs in some jurisdictions. The increased cost of solid waste is exciting both consumers and legislators. Although the sense of urgency in this crisis will diminish in a few years when the political process develops techniques for acquiring new disposal sites, it will resurface in another decade when *those* sites are filled. The cycle of increasing costs and resistance to landfills and incinerators will be resolved at successively higher economic and political costs as long as our economy and population continue to grow.

In this crisis environment, packaging, because it is a highly visible yet poorly understood component of the waste stream, is an attractive target for regulators who are pressed to take action. Packaging regulations will not solve the solid waste problem, however. In fact, ill-advised regulation may harm the environment while imposing excessive costs. Regulations enacted during the intervals of crisis will not disappear when the crisis does.

To prevent ill-advised regulations, industry must be responsive and must also be perceived as responsive. It must work with regulators and legislators each year so that they can take visible and appropriate action on packaging while they are solving the more fundamental problems of landfill and incinerator siting. Such action includes coding of plastic containers for recycling, development of recycling infrastructures, and elimination of toxic metals from inks and pigments in packaging. These steps are more beneficial to the environment and less injurious to the economy than package bans and taxes. In addition to these concrete steps to alleviate the solid waste load, industry must present sound information to the public on the importance of packaging in our society and its impact on the solid waste problem.

Industry must work either directly or through trade associations to educate regulators and legislators and to oppose measures injurious to business and the environment. To be effective in these actions, it must maintain an increased surveillance of state and local legislative activity in this area.

Trade associations are attuned to the issue in varying degrees.

IX. Initiate and support appropriate research and development.

A number of firms are supporting research and development of new methods of recycling and fabricating materials and products from recycled refuse. These ventures, referred to in Chapter 4, should also yield new methodologies and products. Creative partnerships should be sought to research the following:

- More cost-effective collection and sorting.
- Cost-effective methods to recycle multilayer similar and dissimilar materials.

COMMON LEGISLATIVE ACTIONS

Various patterns in legislative initiatives have emerged in several states and municipalities. Some of these have positive environmental impact, some negative.

Several states have programs requiring their local authorities to initiate integrated waste management programs. These can be very helpful but they must be studied closely to make sure there are not hidden pitfalls in the programs.

Several localities have proposed or passed bans on specific package technologies. These are generally inspired by a desire to eliminate materials that are perceived to be environmentally harmful due to their nondegradability or the difficulty in recycling them. These bans are often environmentally harmful, as the substitute material tends to be both heavier and bulkier. The result of such bans is to increase the total amount of solid waste to be handled. *A material should be banned only if the ban will not defeat source reduction while accomplishing the recycling goal.*

Taxes on packaging are a popular proposal for raising revenue for the development of solid waste programs. Such taxes have certain disadvantages. First of all, poor people buy their goods in smaller unit quantities and therefore, on a per dollar basis, are more affected by packaging taxes. This means that packaging taxes are extremely regressive. Second, packaging taxes distort the economics of solid waste. Because packaging makes up only about a third of solid waste, taxing it to pay for the whole problem does not provide the proper incentive to other sources to reduce their impact.

Appendix A

The EPA Agenda for Action

The United States Environmental Protection Agency has issued a report that is required reading for anyone who wishes to understand the solid waste problem in America. The report is *The Solid Waste Dilemma: An Agenda for Action.* It summarizes the problem and sets forth an agenda for action by the EPA, state and local governments, industry, and private citizens. The report includes the data the EPA used to prepare its recommendations.

The document acknowledges the local nature of the solid waste problem and emphasizes that solutions will depend on regional waste types, land use, and demographics. It notes, however, that national trends do exist.

The report focuses on a description of "Integrated Waste Management" systems. The EPA states a goal of reducing the nation's waste 25 percent by 1992, through source reduction and recycling.

To meet this goal, the EPA proposes: (1) to provide a national clearinghouse for information; (2) to provide peer matching for exchange of expertise between levels of government and between government and industry; and (3) to develop a national research agenda for collection of new information and development of new technologies.

The report recommends planning at all levels of government. It gives guidance for such planning and recommends that conferences be organized to facilitate the process.

Regarding source reduction and recycling, the EPA emphasizes the need to reduce the toxicity of waste (particularly lead and cadmium) as well as the total amount of waste. Industrial cooperation in this area is vital. The EPA is planning a corporate recognition program to spur corporate involvement as well as a series of workshops on design for source reduction.

To foster recycling, the EPA recommends that the government implement federal procurement guidelines and recycling of federal waste, and work to stimulate and stabilize markets for secondary materials. It also recommends the formation of a National Recycling Council.

The report states that ash disposal and air emissions from incinerators and landfills deserve continued attention. It recommends that research continue to determine whether bans are necessary.

Appendix B

General Legislative Framework For Environmental Issues

The waste handling business is driven by legislation at the federal, state and local municipal levels. Federal laws (Exhibit B.1) provide the legal framework for other governmental bodies to follow.

Major Themes in State Legislation

Rigorous state and local regulations are taking several forms in relation to packaging. The basic themes or underlying reasons for the legislation are to reduce or eliminate certain kinds of packaging materials—primarily plastic—and to reduce roadside and other litter. The regulators aim to achieve these goals by two principal means: bans and economic incentives.

Outright bans on plastic packaging materials can be either total or selective, but are usually the latter. Bans are gradually phased in over time or started on a specified date.

(continued, page 82)

Exhibit B.1. Key federal legislation governing the waste environment.

Name of Law	Year Passed	Major Features
Clean Air Act	1970; amended 1974, 1977, due to be renewed in 1989	Three air standards to be set by EPA governing ambient air, stationary sources, hazardous air pollutants; states to determine how to meet standards; state/federal partnerships—state to provide permits, enforcement; mobile source standard setting, procedures, enforcement; vehicle emission standards; EPA can take legal action to require cleanup, to be paid for by responsible party.
Clean Water Act	1972; amended 1977, 1980, 1981, 1987	Requires EPA to set quality, pretreatment standards, effluent limitations, and to establish a permit program to regulate pollutant discharges. Minimum performance standards are required but not specific control technologies; covered are point, nonpoint sources; some states authorized to run permit programs. Grant program included to assist in compliance. Toxic pollutant controls increased with amendments, which also create requirements for municipal sewage sludge permitting. State water quality standards identify intended uses for water bodies, then indicate conditions necessary to maintain them.
National Environmental Protection Act (NEPA)	1969	Requires federal agencies to plan their policies and actions in light of environmental consequences; directs preparation of Environmental Impact Statements (EIS) for major federal action significantly affecting the quality of the human environment.
Resource Conservation and Recovery Act (RCRA); reauthorized as Hazardous and Solid Waste Amendments of 1984.	1976; scheduled for reauthorization in 1989.	First comprehensive framework for hazardous wastes management; requires identification, listing, standards for generators, recordkeeping, tracking (in cooperation with U.S. Department of Transportation) performance standards; permits required to control treatment, storage, disposal; requires corrective action at sites with past or currently deposited hazardous wastes. Subtitle D provides for development of environmentally sound disposal methods; regulation of underground storage tanks; protection of groundwater with new landfill standards, restrictions.

Name of Law	Year Passed	Major Features
Comprehensive Environmental Response, Compensation and Liability Act (CERCLA), amended, reauthorized as the Superfund Amendments and Reauthorization Act (SARA) of 1986.	1980, 1986	To provide funds, legal authority for RCRA abandoned hazardous site cleanup; five-year program (to 1991 with SARA); to eliminate most serious threats; Title I deals with release of substances, liability for releases, compensation; Title II deals with taxes on chemical, oil industries to create Trust Fund; taxes on disposal site operators for postclosure expenses; Title III establishes community right-to-know reporting procedures regarding hazardous substances. Directs EPA to establish threshold quantities of hazardous substances; dictates removal, remedial actions to be taken in event of hazardous substance releases on emergency and nonemergency basis.
Safe Drinking Water Act	1974; amended 1976, 1979, 1986	To establish and enforce regulations and standards with states to have major enforcement responsibility; establishes control of underground injection; maximum contaminant levels (legal limits)—not enforceable—for 83 common pollutants. Lead is banned.
Federal Insecticide, Fungicide and Rodenticide Act (FIFRA)	1947; amended 1972, 1975, 1978	EPA regulates use of pesticides and amount of residue to be present in food; Food and Drug Administration (FDA) enforces residue amounts in all food but meat, poultry, and eggs; the latter are monitored by the U.S. Department of Agriculture. EPA registers all pesticides and conditions of their use.
Toxic Substance Control Act (TSCA)	1976	EPA has broad authority over all chemical substances during all phases of their life cycle; pre-manufacture notifications; testing by EPA; control of chemicals with unreasonable risk to health or environment; EPA can create controls for such chemicals as hazardous substances.

Source: *An Introductory Guide to the Statutory Authorities of the United States Environmental Protection Agency*, Region V, April 1988. Document No. EPA.905/9-88-001.

The laws creating the bans often permit exemptions under certain conditions.

Examples of special exemptions may include continued use of existing materials when substitutes meeting similar package specifications cannot be found, when approved materials are not technologically available or feasible, or when approved substitutes are cost-prohibitive. The bans can also be superseded or invalidated by countermanding legislation from a higher governmental level.

Both economic incentives and disincentives may be employed. Taxes are added to products employing certain packaging materials to discourage their use. Taxes collected in this manner are then used to support recycling efforts.

In some states, consumers are required to pay deposits on metal cans, bottles, and other containers to encourage their return. Municipalities such as the borough of Perkasie, Pennsylvania, and High Bridge, New Jersey, require consumers to purchase special stickers or refuse bags. The fee increases with the number used, thereby discouraging the purchase of products with significant packaging.

The federal government leaves a number of areas of authority to state and local governments. Sometimes the federal and state agencies will handle affairs on a partnership basis, as when filing suit against a particular polluter. The granting and enforcement of permits is usually left to state authorities, as in the case of the Clean Air Act. In some instances, states are given this authority because they have taken a more aggressive standard-setting and enforcement role than the federal government.

THE FEDERAL LEVEL

In spite of the growing public awareness, the history of federal involvement in environmental legislation and enforcement has been somewhat checkered. Prior to the 1969 National Environmental Protection Act, most environmentally related regulations were issued by state and local authorities. Federal involvement was dampened during the Reagan years, resulting in political scandals and increasing public discontent. The situation was caused both by budget cutbacks and by the decision to weaken federal regulations and enforcement in order to promote industrial competitiveness. The Bush administration is increasing its emphasis on environmental protection. The EPA is pursuing polluters aggressively with heavy fines and more rigorous enforcement.

The eight legislative examples given above provide a broad framework for air, water, and ground environmental controls dealing with solid and hazardous wastes. Common issues addressed by these laws are:

- Quality standards—such as maximum parts per million permissible amounts of substances to be emitted
- Permitting—to haul or landfill wastes; to emit certain amounts of pollutants; requirements for permits to be granted
- Enforcement of standards
- Funding, such as the $10 billion funding of superfund site cleanup and grant programs to assist in compliance; taxes to create cleanup funds
- Recordkeeping, identification, and tracking of wastes
- Legal authority for cleanup of hazardous wastes

The regulations do dictate minimum quality and performance standards that waste generators must meet. However, they usually do not regulate how, or what technologies must be used to achieve those standards. The federal Environmental Protection Agency has primary authority for regulation of wastes. Its director reports to the President of the United States. Other U.S. agencies that may become involved on certain issues are:

Department of Transportation (DOT)	Monitors and issues permits for the transport of hazardous wastes.
Food and Drug Administration (FDA)	Controls fungicide residue amounts on all foods except meat, poultry, and eggs.

Department of Agriculture   Controls fungicide residue on meat, poultry, and eggs.

As might be expected, these overlapping authorities do create confusion at times. Various states, regional governmental bodies, and the EPA are addressing the need for better coordination.

More than any other document at the federal level, the EPA's "Agenda for Action" (Appendix A) establishes the framework for control of packaging waste.

Appendix C

Solid Waste, Packaging-Related University Programs

Name	Undergraduate	Graduate	Research	Specialties/Programs	Phone
University of California at Los Angeles (UCLA)	Extension Program			Municipal Solid Waste (MSW) Problem Solving, Computerized Information Clearinghouse	213/206-3071
Cornell University (Ithaca, NY)			x	Waste Management Institute Solid, industrial, other wastes	607/255-8674
Florida Institute (Melbourne)	x	x		Practical Solutions to Local Problems	407/768-8000
Georgia Institute of Technology (Atlanta)		x	x	Co-disposal of MSW, hazardous wastes in landfills	404/894-3806
Illinois Institute of Technology (Chicago)		x	x	Recycling, reusing industrial by-products	312/567-3535
University of Illinois (Champaign/Urbana)		x	x	Solid Waste program One of 8 EPA "Centers for Excellence"	217/333-6965
University of Indiana (Bloomington)		x		Solid Waste courses	812/335-9485

University of Michigan (Ann Arbor)	x	Solid Waste Engineering program	313/763-5069
Michigan Technical Institute (Houghton)	x	Solid Waste Management	986/487-2530
University of Nebraska (Lincoln)	x	Solid Waste Management survey course	402/544-2980
Rensselaer Polytechnic Institute (Troy, NY)	x	Solid Waste Research Polymeric recovery from plastics	578/276-6416
Rutgers University (New Brunswick, NJ)	x	Plastic Recycling Research Center: all phases of plastic recycling	201/932-3679
		Cook College: recycling	201/932-9271
		Institute of Packaging Engineering: recycling	201/932-9158
		Environmental Communications Research program: resolving issues through improved communications	201/932-4006

Name	Undergraduate	Graduate	Research	Specialties/Programs	Phone
SUNY (Stony Brook, NY)	x			Certificate program, including recycling, plastic degradation	501/632-8704
Washington State University (Pullman)		x		Solid Waste	509/335-3175
Widener College (Chester, PA)				Extension Program: • Residential recycling • Plastics recycling • Recyclables marketing • Socioeconomic, other factors affecting recycling	215/499-4193
University of Wisconsin (Madison)		x	x	• Core, advanced courses: solid, hazardous waste engineering • Research: recycling	608/262-0493, 608/263-7429

Source: *Waste Age*, December *1988*.

Appendix D

Selected Organizations Dealing with Packaging, Solid Waste Management

Today there are many public and private U.S. organizations working on environmental issues. Below is a select list of groups involved in packaging and solid waste challenges. The list is divided into three categories: public interest groups; industry and trade associations; and government and public organizations.

PUBLIC INTEREST GROUPS

Center for Environmental Education, 202/429-5609: Solid waste management; plastics.
Defenders of Wildlife, 202/659-9510: Solid waste management; packaging. Primary goal is habitat protection. Sup-

ports pro-environment packaging (recyclable and degradable).

Environmental Action Foundation, 202/745-4879: Solid wastes, including recycling, degradable plastics, and packaging.

INFORM Inc., 212/689-4040: Solid waste management.

Keep America Beautiful, 203/323-8987: Waste-handling practices and reducing litter.

Natural Resource Defense Council, 212/727-2700: "Shadow EPA." Waste reduction, solid waste, other topics.

INDUSTRY/TRADE ASSOCIATIONS

Aluminum Association, 202/862-5100: Promotes voluntary aluminum can recycling programs, waste reduction efforts, resource recovery.

American Paper Institute, 212/340-0600: Paper Recycling Committee. Integrated waste management, data collection.

American Recycling Association, 202/785-0550: Aluminum recycling plants. Promotes aluminum recycling.

Can Manufacturing Institute, 202/232-4677: Can manufacturers and suppliers. Favors source reduction, voluntary recycling.

Council for Solid Waste Solutions, 202/382-4610: Promotes plastic recycling.

Council on Plastics and Packaging in the Environment: 202/789-1310. Promotes recycling.

Fiber Box Association, 312/364-9600: Corrugated box manufacturers. Solid waste management.

Flexible Packaging Association, 202/842-3880: Manufacturers, convertors, suppliers, and end-users of flexible packaging. Solid waste management; recycling.

Food Service and Packaging Institute, 202/347-0020: Manufacturers and distributors of food service items and disposable packaging. Solid waste management; integrated waste disposal.

Glass Packaging Institute, 202/887-4850: Glass container manufacturers. Promotes recycling.
Institute for Scrap Recycling Industries, 202/466-4050: Scrap metal industry. Promotes recycling.
National Association of Manufacturers, 202/637-3000.
National Association for Plastic Container Recovery (NAPCOR). 704/357-3250: Promotes curbside and other collection of plastic beverage bottles.
National Environmental Development Association, 202/638-1230: Concentrates on impact of environmental legislation on business, industrial profits.
National Food Processors Association, 202/639-5900: Food processors and suppliers. Solid waste disposal.
National Recycling Coalition, 402/479-3637: Solid waste management: reduction, reuse, recycling. Bringing business and community interests together to promote nationwide plans for waste reduction, recycling programs.
National Soft Drink Association, 202/463-6732: Soft drink bottlers, manufacturers, marketers, suppliers. Solid waste management: recycling, integrated disposal.
National Solid Wastes Management Association, 202/659-4613: Largest group representing solid waste industry. Publishes *Waste Age*. Sponsors Waste Recyclers Council.
The Packaging Coalition for Solid Waste Management, 212/595-9194: Provides education for packaging industry on solid waste issues.
Paperboard Packaging Council, 202/289-4100: Manufacturers of paperboard packaging. Solid waste management; recycling.
Plastics Institute of America, 201/420-5552: Promotes recycling.
Plastics Recycling Foundation, 202/371-5200: Conducts research in plastics recycling.
Polystyrene Packaging Coalition, 202/822-6424: Focuses on polystyrene, recycling.
Pro-Environment Packaging Council, 212/371-2200: Leading companies in the paper and paperboard industry, formed in 1988.

Society of Packaging Professionals (SPHE), 703/620-9380: Promotes recycling, recyclable package design.

Society of the Plastics Industry, 202/371-5200: Parent organization of the Council on Plastics and Packaging in the Environment, and Council for Solid Waste Solutions. Promotes recycling; developed "Voluntary Plastic Container Coding System."

The Steel Can Recycling Institute, 412/922-2772: Formed in 1988 to develop and promote national recycling programs for steel beverage and food containers.

U. S. Chamber of Commerce, 202/463-5533: Represents the broad base of service and industrial businesses at the federal, regulatory level. Solid waste management; integrated disposal.

GOVERNMENT/PUBLIC ORGANIZATIONS

American Legislative Exchange Council, 202/547-4646: Organization of 2,000 state legislators throughout United States. Solid waste management issues developed through National Solid Waste Working Group, a government/industry coalition, that advises on solid waste issues.

Coalition of Northeastern Governors, 202/783-6674: Coalition of nine New England and Mid-Atlantic states. Solid waste management. Has developed a source reduction task force and an executive information exchange on market development for recyclables in Northeast.

National Governors Association, 202/624-5300: Represents governors of fifty states, five territories. Solid hazardous waste management.

National League of Cities, 202/626-3030: Organization of 1,400 cities, towns, and state municipal leagues representing 90 percent of all municipalities with populations greater than 100,000. Solid and hazardous waste management.

U.S. Conference of Mayors, 202/293-7330: Organization of city governments with populations exceeding 30,000. Solid waste management. Umbrella organization for National Resource Recovery Association.

Source: *Waste Age,* January 1989; author files.

Additional Reading

Below are suggested additional sources of information on the subject of packaging and solid waste management:

Bio-Cycle. The JG Press, Inc., Box 351, 18 S. Seventh St., Emmaus, PA 18049. 215/967-4135. An issue of special interest is that of May-June 1988, containing an in-depth report on curbside recycling.

The Management of World Wastes. Channel Communications, Inc., 6255 Barfield Rd., Atlanta, GA 30328. 404/256-9800.

Plastics Recycling: A Strategic Vision. Plastics Recycling Foundation, 1275 K Street N.W., Suite 400, Washington, D.C. 20005. 202/371-5200.

Resource Recovery. National League of Cities Institute, 1301 Pennsylvania Ave., Washington, D.C. 20004. 202/626-3000.

Resource Recycling. Resource Recycling, Inc., 1218 N.W. 21st, Portland, OR 97210, 503/227-1319. Articles of specific interest include:

- Quigley, Jim. "Recycling Frontiers: Glutted Markets and Spurred Demand." March/April 1989.
- Watson, Tom. "Pulp and Paper Mini-Mills Recycle without Fanfare." July 1989.
- Iannazzi, Fred D. "The Economics Are Right for U.S. Mills to Recycle Old Newspapers." July 1989.

- Stauffer, Roberta F. "Energy Savings From Recycling." January/February 1989.
- Ferrand, Trisha L. "Plastics Collection and Sortation." January/February 1989.

Selke, Susan. "Plastics Recycling—Will It Work?" *Journal of Packaging Technology*, October 1987, Technical Publications, Inc., One Lethbridge Plaza, Mahwah, NJ 07430. 201/529-3380.

Packaging industry publications are carrying increasing coverage of packaging-related solid waste issues. These include:

Dairy Foods. Gorman Publishing Co., 8750 W. Bryn Mawr Ave., Chicago, IL 60631. 312/693-3200.

Food & Beverage Marketing. Charleson Publishing Co., P.O. Box 325, 102 Highland Ave., Norwalk, CT 06854.

Food & Drug Packaging. Edgell Communications, Inc., 7500 Old Oak Blvd., Cleveland, OH 44130.

Journal of Packaging Technology. Technical Publications, Inc., One Lethbridge Plaza, Mahwah, NJ 07430. 201/529-3380.

Modern Plastics. McGraw-Hill Inc., 1221 Sixth Ave., New York, NY 10020. 212/512-6242.

Packaging Digest. Delta Communications, Inc., 400 N. Michigan Ave., Chicago, IL 60611. 312/222-2000.

Packaging Magazine. Cahners Publishing Co., 1350 E. Touhy Ave., Des Plaines, IL 60018. The August 1989 issue includes a special in-depth report on "Packaging and the Solid Waste Problem."

Prepared Foods. Gorman Publishing Co., 8750 W. Bryn Mawr Ave., Chicago, IL 60631. 312/693-3200.

Waste Age. National Solid Wastes Management Association. Suite 1000, 1730 Rhode Island Ave., N.W., Washington, D.C. 20036. 202/861-0708. Specific articles, issues of interest include:

- Snow, Darlene. "Plastics Packaging Under Attack." July 1988.

Members of the AMA Packaging Council

Mr. Robert L. Esse
Department Head of Advanced Packaging Technologies
General Mills

Chairman in Charge of the AMA Packaging Council

Dr. Mary A. Amini
Packaging Research Scientist
General Foods USA

Ms. Deborah A. Becker
Associate Director,
Refrigerated Packaging
Kraft, Inc.

Mr. Richard C. Belthoff
Senior Manager,
Packaging Engineering
Warner Lambert Company

Mr. William J. Brown
Manager, Packaging &
    Materials Handling
Primary Glass/Glass Group
PPG Industries, Inc.

Mr. Augustus L. Davis
Director of Package
    Development
Avon Products, Inc.

Mrs. Barbara M. Donnelly,
    C.P.M.
Manager, Packaging
    Procurement, C.P.I.
Johnson & Johnson Consumer
    Products, Inc.

Mr. Van Doran F. Douglass,
    C.P.M.
Manager, Regulatory Services
J. T. Baker Inc.
Division of the Procter &
    Gamble Company

Ms. Jennifer C. Griffin
Advanced Engineer
Corporate Package
    Development
Polaroid Corporation

Dr. Steven W. Gyeszly
Packaging Systems
    Engineering Program
Mechanical Engineering
Texas A & M University

Mr. Thomas J. Hajec
Senior Packaging Engineer
Haworth, Inc.

Mr. Edward A. Hardwidge
Manager, Packaging Research
The Upjohn Company

Mr. Carl C. Hein
Vice President of Packaging
 Technology, RJR Tobacco
R. J. Reynolds Tobacco
 Company

Mr. James H. Houchens
Manager, Corporate
 Packaging Technology
Miles Inc., AB003

Dr. James D. Idol, Jr.
Director, Center for
 Packaging Science &
 Engineering
Rutgers University

Mr. Edmund A. Leonard
Cornell University

Mr. Thomas J. Lowery, CP-P
Director of Engineering
Ethan Allen, Inc.

Mr. William J. Mallin
Director, Product Planning
Purdue-Frederick, Inc.

Mr. Kenney M. Marengo
Packaging Procurement
 Manager
Materials and Logistics
 Department
E. I. du Pont de Nemours &
 Co., Inc.

Mr. Jonathan J. Prinz
President
The Schechter Group, Inc.

Mr. Barry S. Rope
Manager, Corporate
 Packaging Engineering
Corporate Operations
UNISYS Corporation

Mr. Jerry D. Stone
Packaging Advisor
National Classification
 Committee of the Motor
 Carrier Industry

Mr. Robert J. Ten Eyck
Manager, Technical Services
Packaging Engineering
Ecolab, Inc.

Mr. C. A. Tuson
Advisory Engineer
Competence Center for
 Packaging Engineering
IBM Corporation

Mr. Richard J. Vlasich
Packaging and Labeling
 Manager
J. C. Penney Company, Inc.
Lincoln Center II

Ms. Patricia O. Walcott
Research & Development
 Director
M&M/Mars

Mr. David D. Wenberg
Vice President—Materials
 Management
Hillshire Farm & Kahn's Co.

Note: Opinions expressed in this Briefing may differ from the official positions of individual Council members.